Managing
COMPLEX
PROJECTS

Managing
COMPLEX
PROJECTS

Harold Kerzner, Ph.D.

Carl Belack, PMP®

WILEY

John Wiley & Sons, Inc.

INTERNATIONAL
Institute for Learning, Inc.

Copyright © 2010 by International Institute for Learning, Inc., New York. All rights reserved

Published by John Wiley & Sons, Inc., Hoboken, New Jersey
Published simultaneously in Canada

For general information about our other products and services, please contact our Customer Care Department within the United States at (800) 762-2974, outside the United States at (317) 572-3993 or fax (317) 572-4002.

Wiley also publishes its books in a variety of electronic formats. Some content that appears in print may not be available in electronic books. For more information about Wiley products, visit our web site at www.wiley.com.

"PMI," the PMI logo, "OPM3," "PMP," and "PMBOK" are registered marks of Project Management Institute, Inc. For a comprehensive list of PMI marks, contact the PMI Legal Department.

ISBN: 978-0-470-60034-4 (cloth); ISBN 978-0-470-92798-4 (ebk); ISBN 978-0-470-92799-1 (ebk); ISBN 978-0-470-92800-4 (ebk)

Printed in the United States of America

10 9 8 7 6 5 4 3 2 1

CONTENTS

Chapter 3:
SCOPE MANAGEMENT   141

Chapter 4:
TIME MANAGEMENT   161

PREFACE

For more than 50 years, project management has been in use but perhaps not on a worldwide basis. What differentiated companies early on was whether they used project management, not how well they used it. Today, almost every company uses project management, and the differentiation among companies is whether they are simply good at project management or whether they truly excel at project management. The difference between using project management and being good at project management is relatively small, and most companies can become good at project management in a relatively short time period, especially if they have executive-level support. But the difference between being good and excelling at project management is quite large.

For more than three decades, we have become experts in how to manage traditional projects. These traditional projects can be for internal as well as external clients. With these projects, the statement of work is reasonably well defined; the budget and schedule are realistic; reasonable estimating techniques are used, perhaps even estimating databases; and the final target of the project is stationary. We use a project management methodology that has been developed and undergone continuous improvements after use on several projects, and we are able to capture best practices and lessons learned. This traditional project methodology focuses on linear thinking; we follow the well-defined life-cycle phases, and we have forms, templates, checklists, and guidelines for each phase.

Now that we have become good at these traditional projects, we are focusing our attention to the nontraditional or complex projects. The following table shows some of the differences between managing traditional and nontraditional projects:

Traditional Projects	Nontraditional Projects
Time duration of 6–18 months	Time duration can be over several years
The assumptions are not expected to change over the duration of the project	The assumptions can and will change over the project's duration
Technology is known and will not change over the project's duration	Technology will most certainly change
People that started on the project will remain through to completion (the team and the project sponsor)	People that approved the project and are part of the governance may not be there at the project's conclusion
The statement of work is reasonably well-defined	The statement of work is ill-defined and subject to numerous scope changes
The target is stationary	The target may be moving
There are few stakeholders	There are multiple stakeholders

Companies like IBM, Hewlett-Packard, Microsoft, and Siemens are investing heavily to become solution providers and assist clients on a worldwide basis on managing nontraditional, complex projects. Some of the distinguishing characteristics of complex projects, just to name a few, include:

- Working with a large number of stakeholders and partners, all at different levels of project management maturity, and many of whom may not even understand the technology of the project or project management practices

- Dealing with multiple virtual teams located across the world, and where decisions on the project may be made in favor of politics, culture, or religious beliefs

- Starting projects with an ill-defined scope, thereby requiring numerous scope changes throughout the project and, consequently, having a moving target as an end point

- Working with partners and stakeholders that may have limited project management tools and antiquated processes that are incompatible with the project manager's tool kit

- Long-term projects in which the stakeholders may change, new applicable technologies may emerge, and for which funding needs to be justified on a regular basis

- Project in which the stated goals and objectives are not shared by all key stakeholders

For companies to be successful at managing complex projects on a repetitive basis and function as a solution provider, the project management methodology and accompanying tools must be fluid or adaptive. This means that you may need to develop a different project management methodology to interface with each stakeholder given the fact that each stakeholder may have different requirements and expectations, and the fact that most complex projects have long time spans. And while the processes in the *PMBOK® Guide* remain useful on complex projects, it's often necessary to supplement the tool set normally used by project managers employing those processes.

The project manager capability set is necessarily expanded for the management of complex projects. To manage projects with the characteristics noted above, the project manager needs to be able to thrive in and manage an environment of constant change—change in technologies, change in the business and market environments, change in organizational structures and policies, and change among the project's key stakeholders. This requires an increased deftness in the management of what are traditionally known as the "soft skills" of project management—team building, stakeholder management, and leadership, to name a few. There has always been a need for technical credibility and some business knowledge in traditional project management. However, managing complex projects, with their emerging emphasis on returning real business value to both the owner and the contractor, requires an added understanding of the business implications not only of the project itself but also of the project's end product and its value to end users. Finally, the transnational nature of many complex projects requires both political astuteness and cultural sensitivity.

The 4th edition of the *PMBOK® Guide* does an excellent job emphasizing the importance of stakeholder management. Stakeholder management, the first process of the Communications Management knowledge area, may very well be one of the keys to successful management of complex projects. Equally important is the management of project risk, since all of the uncertainties associated with the management of complex projects boils down to risk management. The mastering of the remaining processes of the Communications Management knowledge area, an area of project management in which project managers spend the preponderance of their time, is also a critical success factor in the management of complex projects.

In this book, we first set out to describe project management in terms of its application to, and the differences between, traditional and complex projects. We spend the rest of our time looking at each of the nine knowledge areas of the *PMBOK® Guide* and show how some of the knowledge may have to be applied differently when managing complex projects. The *PMBOK® Guide* is certainly applicable to complex projects, but other factors, such as enterprise environmental factors, may take on a higher degree of importance than they normally would.

HAROLD KERZNER, PH.D.
CARL BELACK, PMP®

ACKNOWLEDGMENTS

Some of the material in this book has been either extracted from or adapted from Harold Kerzner, *Project Management; A Systems Approach to Planning, Scheduling and Controlling*, 10th ed. Hoboken, NJ: John Wiley & Sons, 2009.

Reproduced by permission of Harold Kerzner and John Wiley & Sons, Inc.

We would like to sincerely thank the dedicated people assigned to this project, especially the International Institute for Learning, Inc. (IIL) and John Wiley & Sons, Inc. staff for their patience, professionalism, and guidance during the development of this book.

We would also like to thank E. LaVerne Johnson, Founder, President & CEO, IIL, for her vision and continued support of the project management profession, and Judith W. Umlas, Senior Vice President, Learning Innovations, IIL for their diligence and valuable insight.

In addition, we would like to acknowledge the many project managers whose ideas, thoughts, and observations inspired us to initiate this project.

INTERNATIONAL INSTITUTE FOR LEARNING, INC. (IIL)

With global operating companies all over the world and clients in 200 countries, IIL is a global leader in training, consulting, coaching and customized course development. IIL's core competencies include: Project, Program and Portfolio Management; Business Analysis; Microsoft Project® and Project Server*; Lean Six Sigma; PRINCE2®**; ITIL®; Leadership and Interpersonal Skills. Using their proprietary Many Methods of Learning™, IIL delivers innovative, effective and consistent training solutions through a variety of learning approaches, including Traditional Classroom, Virtual Classroom, simulation training and interactive, on-demand learning. IIL is a PMI® Charter Global Registered Education Provider, a member of PMI's Corporate Council, an Accredited Training Organization for PRINCE2 and ITIL, a Microsoft Gold Certified Partner and an IIBA® Endorsed Education Provider. Now in its twentieth year of doing business, IIL is proud to be the learning solution provider of choice for many top global companies.

---

*Microsoft Project and Microsoft Project Server are registered trademarks of the Microsoft Corporation.
**PRINCE2® is a trademark of the Office of Government Commerce in the United Kingdom and other countries.

Chapter

1

PROJECT MANAGEMENT FRAMEWORK

PROJECT CHARACTERISTICS

- Have a specific objective (which may be unique or one-of-a-kind) to be completed within certain specifications

- Have defined start and end dates

- Have funding limits (if applicable)

- Have quality limits (if applicable)

- Consume human and nonhuman resources (i.e., money, people, equipment)

- Be multifunctional (cut across several functional lines)

We must begin with the definition of a project. Projects are most often unique endeavors that have not been attempted before and might never be attempted again. Projects have specific start and end dates. In some cases, projects may be very similar or identical and repetitive in nature, but those situations would be an exception rather than the norm. Because of the uniqueness of projects and their associated activities, estimating the work required to complete the project may be very difficult and the resulting estimates may not be very reliable. This may create a number of problems and challenges for the functional manager.

Projects have constraints or limitations. Typical constraints include time frames with predetermined milestones, financial limitations, and limitations regarding quality as identified in the specifications. Another typical constraint may be the tolerance for risk and the amount of risk that the project team or owner can accept. There may also be limitations on the quality and skill levels of the resources needed to accomplish the tasks.

Projects consume resources. Resources are defined as human—people providing the labor and support; and nonhuman—equipment, facilities, and money, for example.

Projects are also considered to be multifunctional, which means that projects are integrated and cut across multiple functional areas and business entities. One of the primary roles of the project manager is to manage the integration of project activities. The larger the project, and the greater the number of boundaries to be crossed, the more complex the integration becomes.

THE COMPLEXITY OF DEFINING *COMPLEXITY*

Projects are usually defined as being complex according to one or more of the following elements interacting together:

- Size
- Dollar value
- Uncertain requirements
- Uncertain scope
- Uncertain deliverables
- Complex interactions
- Uncertain credentials of labor pool
- Geographic separation across multiple time zones
- Other factors

Complex projects differ from traditional projects for a multitude of reasons, many of which are shown in the following feature. There are numerous definitions of a complex project. The projects that you manage within your own company can be regarded as a complex project if the scope is large and the statement of work only partially complete.

Some people believe that research and development (R&D) projects are always complex because, if you can lay out a plan for R&D, then you probably do not have R&D. R&D is when you are not 100 percent sure where you are heading, you do not know what it will cost, and you do not know when you will get there.

Complexity can also be defined according to the number of interactions that must take place for the work to be executed. The greater the number of functional units that must interact, the harder it is to perform the integration. The situation becomes more difficult if the functional units are dispersed across the globe and if cultural differences makes integration difficult.

Complexity can also be defined according to size and length. The larger the project in scope and cost, and the greater the time frame, the more likely it is that scope changes will occur affecting the budget and schedule. Large, complex projects tend to have large cost overruns and schedule slippages. Good examples of this are Denver International Airport, the Chunnel between England and France, and the "Big Dig" in Boston.

COMPONENTS OF COMPLEX PROJECTS

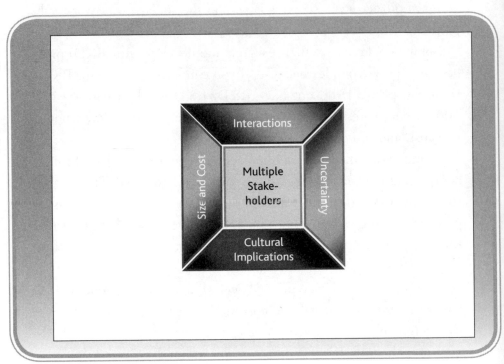

For the purposes of this book, we will consider complex projects to be defined according to the five elements shown in the preceding feature:

- *Size and cost.* According to size, we shall assume that this project is possibly one of the largest and most costly projects that you have ever worked on. The budget could be in hundreds of millions or, if your company works on projects up to $5 million, then this project might be $20 million. Furthermore, the project is being accomplished for a client external to your company.

- *Interactions.* You must interface with several subcontractors or suppliers, and many of them may be in different time zones. You are most likely using a virtual team concept for all or part of the people you must interface with.

- *Cultural implications.* Because some or all of your team members may come from various locations around the globe, cultural differences can have a severe effect on the management of the project.

- *Uncertainty.* This project is unlike any other project you have managed, and there is a great deal of uncertainty. The uncertainty deals with not only the scope and the deliverables, but also with the size of the project team and the cultural differences.

- *Stakeholders.* There are several stakeholders that you must interface with, and getting them all to agree on the scope, the deliverables, and the approval of change requests will be difficult. Stakeholders may have their own agendas for the project, and each stakeholder may have funded part of the project.

THE TRIPLE CONSTRAINT

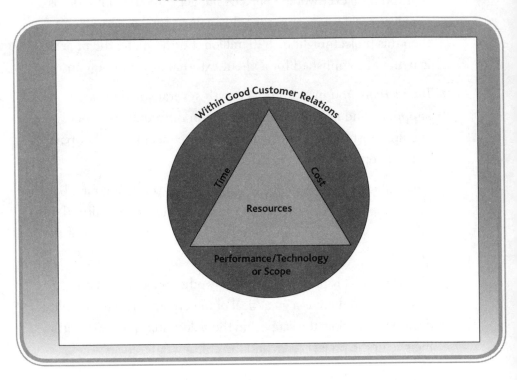

Project management is an attempt to improve efficiency and effectiveness in the use of resources by getting work to flow multidirectionally through an organization. This holds true for both traditional projects and complex projects. Initially, this might seem easy to accomplish, but there are typically a number of constraints imposed on a project. The most common constraints are time, cost, and performance (also referred to as scope or quality), known as the triple constraints.[1]

From an executive management perspective, the preceding feature is the goal of project management, namely, meeting the triple constraints of time, cost, and performance while maintaining good customer relations. Unfortunately, because most projects have some unique characteristics, highly accurate estimates may not be possible, and trade-offs among the triple constraints may be necessary. Executive management and functional management must be involved in almost all trade-off discussions to ensure that the final decision is made in the best interest of both the project and the company. If multiple stakeholders are involved, as there are on complex projects, then agreement from all of the stakeholders may be necessary. Project managers may possess sufficient knowledge for some technical decision making, but may not have sufficient business or technical knowledge to adequately determine the best course of action to address interests of the company as well as the project.

The preceding feature shows that resources are consumed on a project. Typical traditional resources include money, manpower, information, equipment, facilities, and materials. Assuming that the project manager and functional manager are separate roles assigned to different people, the resources are generally administratively under the control of the functional managers. The project managers must

---

[1]Please note that in the *PMBOK® Guide*—Fourth Edition, the triple constraints have been replaced by the concept of "competing demands" or "competing constraints." The new demands/constraints add risk, resources, and quality to the original set.

therefore negotiate with the functional managers for some degree of control over these resources. It is not uncommon for project managers to have minimal or no direct control over project resources and to rely heavily on the functional managers for resource-related issues. The resources may be in a solid line type of reporting relationship to their functional manager and dotted line or indirect reporting to the project manager. The solid-dotted line relationship can become quite difficult to manage if the resources are under the control of functional managers geographically separated from the project manager.

Some people argue that project managers have direct control over all budgets associated with a project. The truth of the matter is that project managers have the right to open and close charge numbers or cost accounts for a project. But once the charge numbers are opened, the team members performing the work and their respective functional managers are actually in control of how the money is being spent as long as the charge number limits are not exceeded. With geographically dispersed teams, the problem of monitoring and controlling funds can create monumental headaches. Currency exchange rates also add to the complexity.

SECONDARY SUCCESS FACTORS

Secondary Factors

- Customer reference
- Commercialization
- Follow-on work
- Financial success
- Technical superiority
- Strategic alignment
- Regulatory agency relationships
- Health and safety
- Environmental protection
- Corporate reputation
- Employee alignment
- Ethical conduct (Sarbanes-Oxley law)

In the previous features, we discussed that time, cost, and performance were the primary components to the triple constraint. Project success is usually measured by how well we perform within the triple constraint. While that is true, there are secondary constraints that can be of greater importance to stakeholders than the primary constraints. As an example, a company agreed to execute a contract for a client at a contract price that was 40 percent below their own cost of doing the work. When asked why they bid on the contract at such a low price and knew full well that they would be losing money, an executive said: "We are doing this only once. We need to the client's name on our resume of clients that we have serviced." In this case, the contractor's definition of success was customer reference.

In another example, the R&D group of a manufacturer of paint products stated that their definition of success was measured by product commercialization. Any R&D project that eventually gets commercialized is viewed as a success. While this definition seems plausible, there may be a problem if marketing and sales cannot find customers for the product. In other words, we can have project success but product/program failure. It is better if both project and program success are achieved.

In a third example, an aerospace company underbid the initial contract to develop a complex product for the Department of Defense. When asked why the R&D effort was bid at a loss, the company responded that they would make up the difference when they were awarded the follow-on contract. In this case, success was measured by the amount of work to be received in the future.

OTHER SUCCESS FACTORS

Other Factors

- With minimum or mutually agreed upon scope changes
- Without disturbing the normal flow of work within the business
- Without changing the corporate culture
- Without a disruption to organizational governance

There are many components of project success. Most components of success involve the deliverables provided at the end of the project. However, for large, possibly long-term complex projects, there can also exist components of success related to changes that occurred in the company in the way the project was executed. On complex projects with multiple stakeholders and possibly several contractors, each company involved in the project can be impacted differently.

First, complex projects have complex scope change approval processes. In an ideal situation, all stakeholders will be in agreement with the scope changes. But if some stakeholders are not in agreement with the scope changes, then the project may have an impact on the way the company does business. This could easily disturb the normal flow of work in a company. As an example, in one company, the approval of a scope change could mandate that the company assign their best employees to the project. This could create a problem if the employees must be removed from other assignments that are critical to the ongoing success of the company.

Another example could involve the corporate culture. Some cultures are heavily oriented around power and authority relationships and may not want to support virtual teams or empower project teams. The approval of certain scope changes could require that the team perform in a manner different from the existing culture. This could also cause a dramatic change in the governance structure of a company.

THE MODIFIED TRIPLE CONSTRAINT

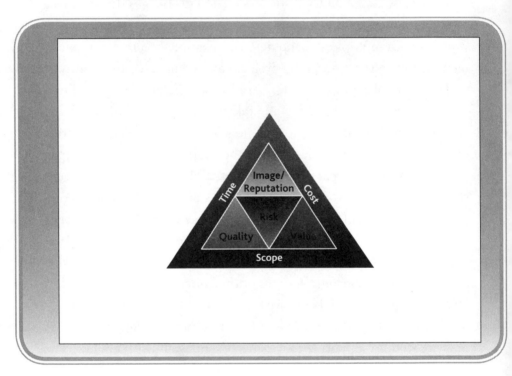

The preceding feature shows the modified triple constraint. The traditional triple constraint that has been used for decades includes time, cost, and scope. Some practitioners prefer to use performance instead of scope, where performance can be scope, quality, or technology. However, for complex projects involving multiple stakeholders as discussed previously, there can be more than three constraints that are considered to be important.

For complex projects, quality, risk, image/reputation, and value can carry a great deal of importance. But the exact degree of importance can vary from stakeholder to stakeholder and from country to country. As an example, a project manager was given an assignment to manage a project from the construction of a large hospital in a developing nation. The project manager's focus was on quality, whereas the host country's priority was simply having the hospital built regardless of cost overruns and schedule slippages. The people would be happy with a hospital, and excessive quality was not important to them.

In some host countries, the project's risk is extremely important, especially if the failure of the project can damage the host country's image or reputation. Risks and politics may go hand in hand in some host countries to the point where the early cancellation of a project may be necessary rather than incurring added risks that could damage one's image.

Some people define value and what the quality is worth. Value may be seen as being more important than cost or schedule. In some cases, such as Denver International Airport and the Opera House in Sydney, Australia, the focus on value allowed for cost overruns and schedule slippages. This is common on large, complex projects.

PRIORITIZATION OF CONSTRAINTS

Over the life cycle of a project, the prioritization of the constraints can change. As an example:

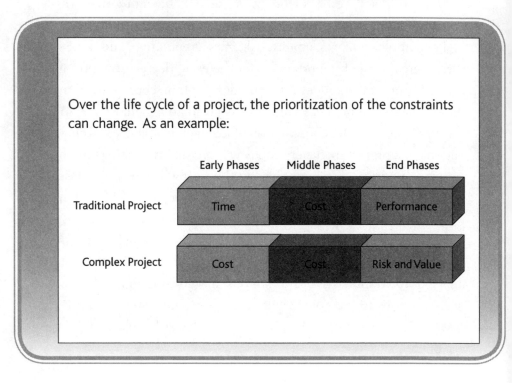

We have talked thus far about the fact that the constraints on a project can be prioritized differently by each of the stakeholders and from project to project. In some companies, the prioritization is almost always the same. For example, years ago Walt Disney had six constraints on the projects involving the development of new attractions at their theme parks: time, cost quality, aesthetic value, safety and scope. The three constraints of safety, aesthetic value, and quality were considered "untouchable" constraints, never to be deviated from. If trade-offs had to take place, they were always on time, cost, and scope.

As can be seen in the preceding feature, the relative importance of each constraint can change from life-cycle phase to life-cycle phase. As an example, in a traditional project, time is critical when planning the project to make sure that we can meet the customer's end-date expectations. In the middle phases, where most of the money is spent, cost becomes important. As we approach the end of the project, performance takes center stage.

For nontraditional or highly complex projects, cost is an issue until we approach the end of the project. At this point, risk and final value become important. However, this is just an example. In large, complex projects, the priorities in each life-cycle phase can change based on stakeholder interests and needs.

TYPES OF PROJECT RESOURCES

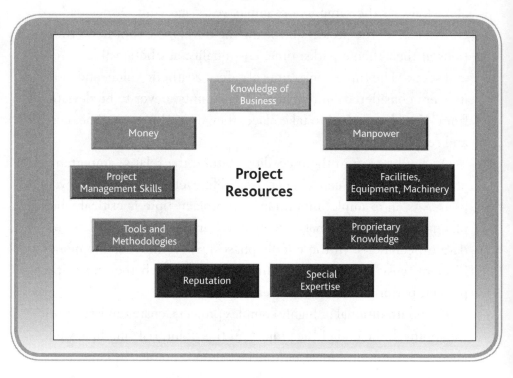

The preceding feature shows the various project resources that project managers may or may not have under their direct control. On complex projects, very few of these are directly under the control of the project manager. Some of these resources require additional comment.

- *Money.* Once budgets are established and charge numbers are opened, project managers focus more on project monitoring of the budget rather than management of the budget. Once the charge numbers are opened, the performers or workers and their respective line managers control how the budgets for each work package will be used. This can be a severe problem for the project manager if the work is being accomplished at a location geographically distant from the project manager.

- *Resources.* Resources are usually "owned" by the functional managers and may be directly controlled by the functional managers for the duration of the project. Also, even though the employees are assigned to a project team, functional managers may not authorize them to make decisions without review and approval of the functional managers.

- *Business knowledge.* Project managers are expected to make business decisions as well as project decisions. This is why executives must become involved with projects and interface with project managers to provide project managers with the necessary business information for decision making.

SKILL SET

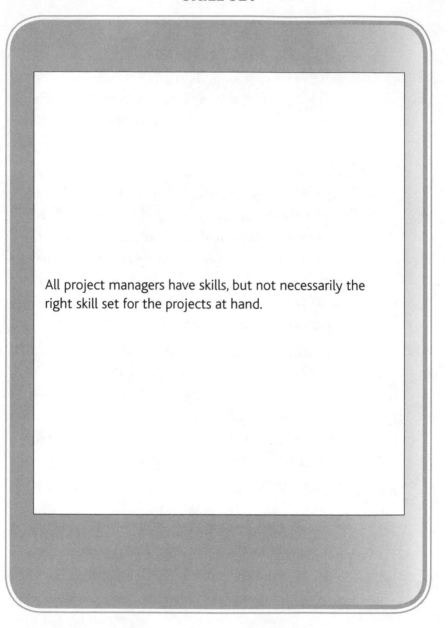

All project managers have skills, but not necessarily the right skill set for the projects at hand.

All project managers have skills, but not all project managers will have the right skills for the right jobs. For projects internal to a company, it may be possible to develop a company-specific universal skill set or company-specific body of knowledge. Specific training courses can be established to support company-based knowledge requirements.

For complex projects with a multitude of stakeholders, all from different countries with different cultures, finding the perfect project manager may be an impossible task. Today, we are in the infancy stages of understanding complex projects and being able to determine the ideal skill set for managing complex projects. We must remember that project management existed for more than three decades before we created the first *Guide to the Project Management Body of Knowledge (PMBOK® Guide)*, and even now with the fourth edition of the *PMBOK® Guide*, it is still referred to as a "guide."

We can, however, conclude that there are certain skills required to manage complex projects. Some additional skills might include how to manage virtual teams, understanding cultural differences; managing multiple stakeholders, each of whom may have a different agenda; and understanding the impact of politics on project management.

THREE CRITICAL REQUIREMENTS

The three critical requirements for successful execution of a complex project include:

■ Clear understanding of the goals and objectives

■ User involvement from cradle to grave

■ Clear governance

I n the previous features, we discussed the importance of understanding the environment in which the complex project will be
executed in order to determine the skills needed by the project
manager. Although there are several factors that can have a major
influence on the project environment, three of these are identified in
the preceding features. With multiple stakeholders and possible cultural
barriers, it is important that the project manager and all stakeholders
have a unified agreement and understanding of the project's goals
and objectives.

Cradle-to-grave user involvement in complex projects is essential.
What is unfortunate is that user involvement can change based on
politics and the length of the project. It is not always possible to
have the same user community attached to the project from beginning to end. Promotions, changes in power and authority positions
due to political elections, and retirements can cause a shift in user
involvement.

Governance is the process of decision making. On large complex
projects, governance will appear in the hands of the many rather
than in the hands of the few. Each stakeholder will either expect or
demand to be part of all critical decisions on the project. The channels for governance must be clearly defined at the beginning of the
project, possibly before the project manager is assigned. Changes in
governance, which is expected the longer the project takes, can have
a serious impact on the way the project is managed.

PROBLEM IDENTIFICATION AND SOLUTION

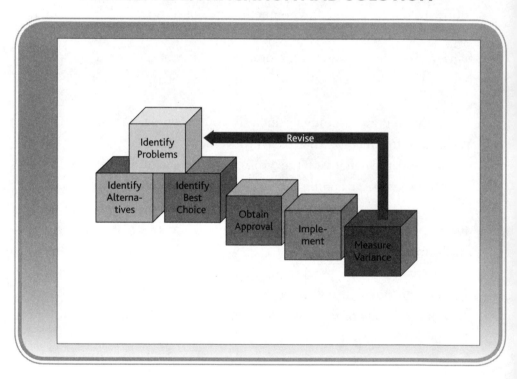

All problems have solutions, but not all solutions are good or even practical. Also, the solution to some problems is more costly than other solutions. The preceding feature illustrates a simplistic approach to problem solving. This approach applies to traditional as well as complex projects, but is more difficult to implement with complex projects.

Identifying a problem is usually easy. Identifying alternatives may require the involvement of many stakeholders, and each stakeholder may have a different view of the actual problem and the possible alternatives. To complicate matters, some host countries have very long decision-making cycles even for the identification of the problem as well as for the selection of the best alternative. Each stakeholder may select an alternative that is in the best interest of a particular stakeholder rather than in the best interest of the project.

Obtaining approval can take just as long, especially if the solution requires that additional capital be raised and if politics take an active role. In some emerging countries, every complex project may require the signature of all of the ministers and senior leaders. Decisions may be based on politics and religion as well.

THE "TRADITIONAL" PROJECT

- The project has a time duration of 6 to 18 months.
- The assumptions are not expected to change over the duration of the project.
- Technology is known and will not change over the duration of the project.
- People who start on the project will remain through to completion (the team and the sponsor).
- The statement of work is reasonably well defined.
- The target is stationary.

In the past several features, we discussed both traditional and complex projects. We will now show the differences between them. The traditional project that most people manage is usually less than 18 months in duration. In some companies, the traditional project might be six months or less. The length of the project is usually dependent on the industry. In the auto industry, for example, a traditional project is three years.

With projects that are 18 months or less, we assume that technology is known with some degree of assuredness and will undergo little change over the life of the project. The same holds true for the assumptions. We tend to believe that the assumptions made at the beginning of the project will remain intact for the duration of the project unless a crisis occurs.

People who are assigned to the project will most likely stay on board the project from beginning to end. The people may be full time or part time. This includes the project sponsor as well as the team members.

Because the project is 18 months or less, the statement of work is usually reasonably well defined and the project plan is based on reasonably well understood and proven estimates. Cost overruns and schedule slippages can occur, but not to the degree that they will happen on complex projects. The objectives to the project, as well as critical dates, are reasonably stationary and not expected to change unless a crisis occurs.

THE "NONTRADITIONAL" (COMPLEX) PROJECT

- Time duration can be over several years.
- The assumptions can and will change over the duration of the project.
- Technology will change over the duration of the project.
- People who approved the project (and are part of the governance) may not be there at completion.
- The statement of work is ill defined and subject to numerous changes.
- The target may be moving.

The complexities of nontraditional projects seem to be driven by time and cost. Complex projects may be as long as 10 years or even longer. Because of the long time duration, the assumptions made at the initiation of the project will most likely not be valid at the end of the project. The assumptions will have to be revalidated throughout the project.

Likewise, technology can be expected to change throughout the project. Changes in technology can create significant and costly scope changes to the point where the final deliverable does not resemble the initial planned deliverable.

People on the governance committee and in decision-making roles most likely are senior people and may be close to retirement. Based on the actual length of the project, the governance structure can be expected to change through the project if the project is 10 years or longer in duration.

Because of scope changes, the statement of work may undergo several revisions over the life cycle of the project. New governance groups and new stakeholders can have their own hidden agendas and demand that the scope be changed or else they might even cancel their financial participation in the project. Finally, whenever you have a long-term complex project where continuous scope changes are expected, the final target may be moving. In other words, the project plan must be constructed to hit a moving target.

WHY TRADITIONAL PROJECT MANAGEMENT MUST CHANGE

New projects have become:

- Highly complex and with greater acceptance of risks that may not be fully understood during project approval

- More uncertain in the outcomes of the projects with no guarantee of value at the end

- Pressed for speed-to-market irrespective of the risks

Traditional project management works well when the direction of the project is clearly understood, the target is stationary, the scope is clearly defined, everyone agrees on the objectives and expectations, the risks are considered low and well understood, and there exists a high probability of project success. But for companies that wish to be innovative and become market leaders rather than market followers, the type of projects approved can be fuzzy and not follow these criteria. This is especially true for complex projects.

More and more projects are highly complex and may require a technical breakthrough. In addition, the risks in achieving the breakthrough are high, and we have no guarantee that we will be successful and that the expected value at the end of the project will be achieved. If a market leadership position is desired, then the projects are further complicated by the requirement to compress the schedule further for an early introduction into the marketplace.

Today's projects are not necessarily as well defined and understood as the traditional projects of the past. As a result, the traditional theories of project management may not work well on these new, complex types of projects. We may need to change the way we manage and make decisions on these projects. Business decisions may very well override technical decisions on projects.

The statement of work (SOW) is:

■ Not always well defined, especially on long-term projects

■ Based on possibly flawed, irrational, or unrealistic assumptions

■ Inconsiderate of unknown and rapidly changing economic and environmental conditions

■ Based on a stationary rather than moving target for final business value

As projects become more complex, the statements of work (SOWs) become less well defined and possibly ill defined. With all SOWs, assumptions are made. On long-term projects, realistic assumptions about politics, environmental conditions, and the economy are almost impossible to make. In such cases, the value achieved in the deliverable can be expected to become more important. Also, the achieved value may not have been fully understood initially, and may have changed over the life of the project. Therefore, the final value of the project may be a moving target rather than a stationary target, and we may have to accept a final value that is quite different from our initial expectations. The longer the project, the greater the chance that the final result will be significantly different than our initial expectations.

Given our premise that project managers are now more actively involved in the business, we must track the assumptions the same way that we track budgets and schedules. If the assumptions are wrong or no longer valid, then we may need to either change the SOW or even consider canceling the project. We should also track the expected value at the end of the project because unacceptable changes in the final value may be another reason for project cancellation.

The management cost and control systems (enterprise project management methodologies [EPM]) focus on:

■ An ideal situation (as in the *PMBOK® Guide*)

■ Theories rather than the understanding of the work flow

■ Inflexible processes

■ Periodically reporting time at completion and cost at completion, but not value (or benefits) at completion

■ Project continuation rather than canceling projects with limited or no value

M ost companies either have or are in the process of developing an enterprise project management (EPM) methodology. EPM systems are usually rigid processes designed around policies and procedures, and work efficiently when the SOW is well defined. But with the new type of projects expected over the next decade, these rigid and inflexible processes may be more of a hindrance.

EPM systems must become more flexible in order to satisfy business needs. The criteria for good systems will lean toward forms, guidelines, templates, and checklists rather than policies and procedures. Project managers will be given more flexibility in order to make decisions necessary to satisfy the business needs of the project. The situation is further complicated in that all active stakeholders may need to use the methodology and having multiple methodologies on the same project is never a good idea. Some host countries may be quite knowledgeable in project management, whereas others may have just cursory knowledge.

In the future, the assumption that the original plan is correct may be a poor assumption. As the project's business needs change, the need to change the plan will also be evident. Also, decision making based entirely on the triple constraint, with little regard for the final value of the project, may be a poor decision.

Simply stated, today's view of project management is quite different than the views in the past, and this is partially the result of having recognized the incremental benefits of project management over the past two decades.

TRADITIONAL VERSUS COMPLEX PROJECTS

Managing Traditional Projects	Managing Nontraditional Projects
Single-person sponsorship	Governance by committee
Possibly a single stakeholder	Multiple stakeholders
Project decision making	Both project and business decision making
Inflexible project management methodology	Flexible or "fluid" project management methodology
Periodic reporting	Real-time reporting
Success is defined by the triple constraint	Success is defined by the triple constraint and business value
Key performance indicators (KPIs) are derived from earned value measurement (EVM)	Unique value-driven KPIs can exist on every project

We can now summarize some of the differences between managing traditional versus complex projects. Perhaps the primary difference is with whom the project manager must interface on a daily basis. With traditional projects, the project manager interfaces with the sponsor and the client, both of whom may be the only governance on the project. With complex projects, governance is by committee, and there can be multiple stakeholders whose concerns need to be addressed.

With complex projects, the project manager needs a fluid or flexible project management methodology capable of interfacing with multiple stakeholders. The methodology may need to be more aligned with business processes than with project management processes since the project manager may need to make business decisions as well as project decisions. Complex projects seem to be dictated more by business decisions than by pure project decisions.

Complex projects are driven more by the project's end value than by the triple constraint. Complex projects tend to take longer than anticipated and cost more than originally budgeted in order to guarantee that the final result will have the value desired by the customers and stakeholders. Simply stated, complex projects tend to be value driven rather than driven by the triple constraint. The reason is simple: completing a project within the triple constraint is not necessarily success if the value is not there at the conclusion of the project.

THE NEED FOR "VALUE" AS A DRIVER

Factors promoting value-driven project management include:

- Identifying the value of business opportunities that do not yet exist
- Identifying better ways of selecting projects with the greatest potential value
- Identifying better ways of measuring the value of projects once they begin and/or end
- Making better decisions in turbulent and highly dynamic markets
- Measuring value has become a competitive necessity
- Implementing client-value programs

In the previous features, we stated that the criteria for project success must have a value component. Projects are not approved or funded based on the triple constraint. Rather, they are selected based on the value that is expected at the end of the project. Simply stated, complex projects appear to be value driven rather than being driven by time or cost.

We are just beginning to find ways to measure value on projects. Traditional forecast reports provide information on the time and cost expected at the completion of the project. This data can be calculated from extrapolation of trends or formulas. Unfortunately, this data may not be sufficient to provide management with the necessary information to make effective business decisions and to decide whether to continue on with the project or consider termination based on the value expected at the end. Most earned value measurement (EVM) systems in use today do not report value at completion of expected benefits at completion, probably because there are no standard formulas for them.

The benefits and value at completion must be calculated periodically throughout the life cycle of the project. However, based on which life-cycle phase a project is in, there may be insufficient data to perform the calculation quantitatively. In such cases, a qualitative assessment of benefits and value at completion may be necessary, assuming, of course, that information exists to support the assessment. Expected benefits and value are more appropriate for business decision making and usually provide a strong basis for continuation or cancellation of the project.

THE BENEFITS OF "VALUE" AS A DRIVER

Value-driven project management leads to:

- Better decision making, especially when considering nonfinancial (intangible) benefits

- Better analysis of options, especially when considering scope changes and trade-offs

- Better alignment of projects to corporate objectives during business case development

- Easier to get stakeholder consensus on value than on just the triple constraint

- Better persuasive and defendable justification for funding during portfolio project selection activities

- Expectation of an increase of 30 percent or higher in total portfolio value, but can be industry-specific

There are significant benefits to using the value component of success as a driver. Value can be measured either quantitatively or qualitatively and provides valuable information at the gate reviews for the continuation of the project, redirection of the project, or simply project termination. Project trade-offs are easier to perform if the decisions are based on value rather than time or cost.

Project stakeholders can be expected to argue over time and cost decisions, but it is usually easier to get consensus when discussing value. Also, during project selection and approval, it is easier to get authorization for funding when the decisions are based on value.

There must exist some form of reevaluation or reexamination process to determine if we are still on track to produce the value expected at completion. However, there may be roadblocks that discourage a reexamination process. As an example, an executive may have funded a "pet" project and may be afraid of the realities that would be discovered during the reexamination process.

Reexamination processes need not be accomplished at the same time as the end-of-phase review meetings that are part of an EPM methodology. They may be accomplished monthly, based on availability of information, at the discretion of specific stakeholder requests, or when a significant change occurs in the political, business, or economic environment.

ELEMENTS OF COMPLEXITY

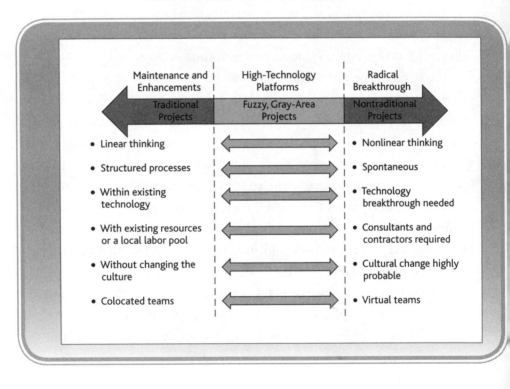

The preceding feature is a summary of what we have discussed thus far about the differences in managing traditional projects. It is important to emphasize that not all projects are easily classified as traditional or complex. There is no fine line between them. Rather, it is a gray area where we have some fuzzy projects that can go either way and be classified as complex or traditional.

As an example, small complex projects are generally managed by co-located teams where all of the team members are removed from their line organization and physically reside in the same location, usually under the supervision of the project manager. This type of project generally does not fall under our definition of a complex project, but for the company needing to perform the work, it could be seen as a complex project.

Most complex projects maximize the use of virtual teams. It is quite uncommon for a company to work on a complex project and discover that all of the employees needed for the project already reside within the company. Virtual project teams allow a company to tap into highly qualified resources elsewhere in the world. The difficulty is that the project manager may not physically interface with these people but may need to rely heavily upon conference calls, web-based meetings, and other media. Superior communication skills are needed for virtual project management teams to be successful.

TYPES OF VIRTUAL TEAMS

Types of Teams	Description
Network	Team membership is diffused and fluid; members come and go as needed. Team lacks clear boundaries with the organization.
Parallel	Team has clear boundaries and distinct membership. Team works in short term to develop recommendations for an improvement in a process or system.
Project or Product Development	Team has fluid membership, clear boundaries and a defined customer, technical requirement and output. Longer-term team task is nonroutine, and team has decision-making authority.
Work, Functional, or Production	Team has distinct membership and clear boundaries. Members perform regular and ongoing work, usually in one functional area.
Service	Team has distinct membership and supports ongoing customer, network activity.
Management	Team has distinct membership and works on a regular basis to lead corporate activities.
Action	Team deals with immediate action, usually in an emergency situation. Membership may be fluid or distinct.

Adapted from Deborah L. Duarte and Nancy Tennant Snyder, *Mastering Virtual Teams*, 3rd ed. Hoboken, NJ: John Wiley & Sons, 2006, p. 9.

It is not our intent to leave the reader with the impression that all complex projects require virtual teams. As can be seen from the above slide, there are numerous applications to virtual teams and therefore several types of virtual teams. Reading through the description of the types of teams, we can identify some critical issues with virtual teams that may affect complex projects:

- Parts of the virtual team may not feel any membership with the project team.

- Each part of the virtual team may possess their own unique project management tools and methodology, and they may not be compatible with the project manager's methodology.

- Loyalty of virtual team members is always a challenge.

- Each portion of the virtual team may have its own governance structure for decision making, and each may have an elongated process for decision making.

- In time of crisis, decision making may be slow.

- Virtual team members may have other duties in their parent company that are a higher priority than your project.

- Virtual team members may be working on multiple virtual teams.

VIRTUAL TEAM COMPETENCIES

- Project management techniques

- Networking across functional, hierarchical, and organizational boundaries

- Using electronic communication and collaboration technology effectively

- Setting personal boundaries and being assertive about being included

- Managing one's time and one's career

- Working across cultural and functional boundaries

- A high level of interpersonal awareness

Adapted from Deborah L. Duarte and Nancy Tennant Snyder, *Mastering Virtual Teams*, 3rd ed. Hoboken, NJ: John Wiley & Sons, 2006, p. 23.

The preceding feature shows some of the additional challenges facing virtual teams. First, not all virtual team members understand project management, nor do they all have the same project management tools. This may be particularly true if part of the team resides in an emerging country.

Not all virtual teams understand how to communicate across organizational boundaries or continents. In some countries, virtual team members must follow the hierarchical chain of command for all communication, even though they are told that they are part of a complex project team. To complicate matters, not all virtual project team members will possess the same technology for communications.

The *PMBOK® Guide* encourages project managers to take an active role in helping project team members become better employees in hopes of achieving rewards, promotion, or advancement opportunities. With virtual project teams, this may be quite a challenge for the project manager. The project manager may never physically see the team members, know whether they are being assisted by their line manager in the performance of their project responsibilities, or know if they are performing up to their capacity. Without physical interfacing, the project manager's participation in a wage and salary administration program is meaningless.

VIRTUAL TEAM MYTHS

- Myth 1: Virtual team members don't need attention.

- Myth 2: The added complexity of using technology to mediate communication and collaboration over time, distance, and organization is greatly exaggerated.

- Myth 3: The leader of a cross-functional virtual team needs to speak several languages, have lived in other countries, and have worked in different functions.

- Myth 4: When you can't see people on a regular basis, it is difficult to help them with current assignments and career progression.

- Myth 5: Building trust is unimportant in virtual teams.

- Myth 6: Networking matters less in a virtual environment; it is only about results.

- Myth 7: Every aspect of virtual teams should be planned, organized, and controlled so that there are no surprises.

Adapted from Deborah L. Duarte and Nancy Tennant Snyder, *Mastering Virtual Teams*, 3rd ed. Hoboken, NJ: John Wiley & Sons, 2006, pp. 76–87.

All too often, virtual teams are formed and the team members have a relatively poor understanding of how virtual teams should function. This occurs because management cannot or does not want to invest in training related to virtual teams. The result is that people end up with myths concerning virtual teams. The preceding feature illustrates some of the myths.

Some project managers erroneously believe that helping team members become better workers does not apply to virtual team members who are remotely located. Project managers should be responsible to help all team members perform to their limits.

Not every aspect of a virtual team can be planned for. Each virtual team can be impacted by their company's culture, politics, governance structure, ability to take risks, and available technology for project communications. Most of the time, virtual team members neglect to inform the project manager about these complexities. These complexities usually become apparent when problems occur and decisions must be made. Project managers need to be aware of the risks associated with managing virtual teams and take steps to avoid or mitigate these risks whenever possible.

CUSTOMER RFP REQUIREMENTS

◼ Contractors must have Project Management Professionals (PMP®s).

◼ Contractors must have an EPM system, and it may have to be qualified or approved by the client.

◼ Contractors must capture best practices and share intellectual property with the client.

◼ Contractors must identify a reasonable maturity level in project management.

The growth of project management is heavily customer driven. By this we mean that customers possess the ability to pressure the suppliers to make improvements to their project management capability in order to win the contract. This pressure for improvement appears in the customer's request for proposal (RFP). As an example, customers are now demanding that suppliers identify in their proposal the number of Project Management Professionals (PMP®s) they have in their company and also identify which PMP® will be managing the project.

Customers are requiring that the supplier clearly identify their EPM methodology and its capability. This is a necessity because the customer expects to be interfacing with the methodology. The supplier may also be informed that the customer must certify that the methodology meets the customer's standards of performance. The alternative would be for the supplier to agree to use the customer's methodology.

Historically, at the end of a project, customers were pleased simply to receive the required deliverables from the project. The supplier would walk away with all of the project management intellectual property that was paid for by the customer. Now the customers are asking for all of the project management best practices and lessons learned that the customer funded.

Customers are now demanding that suppliers identify their maturity level in project management. There are several project management maturity models that can be used to do this. Suppliers can usually select the model most appropriate for their use.

THE NEED FOR BUSINESS SOLUTION PARTNERS

- Not all companies have the ability to manage complexity.

- Solution providers must learn while managing the project.

- Solution providers can bring years of history to the table.

- Solution providers have a greater understanding of cultural change and the ability to work within almost any culture.

Very few companies have sufficient resources such that they can manage a large, complex project by themselves without seeking out external support. As such, contractors are often hired to provide turnkey solutions to complex projects.

Experienced contractors can bring years of experience to the table, as well as an abundance of lessons learned on other project and best practices. These types of contractors pride themselves on being solution providers rather than just contractors. They will promise you a solution to your business problems.

Solution providers generally have sufficiently more project management intellectual capital than the ultimate customer. Solution providers are also more knowledgeable in the use of virtual teams and working within a multitude of different cultures. Solution providers also have the experience in how to accelerate decision-making processes.

A project, by definition, is a unique endeavor that may never have been attempted in the past. As such, managing this type of complex project is a learning experience for both the customer and the solution providers. Both parties must be willing to learn from the successes and failures on a project. It is wishful thinking to believe that new, complex projects will always go as planned without any mistakes or partial failures.

"ENGAGEMENT" EXPECTATIONS

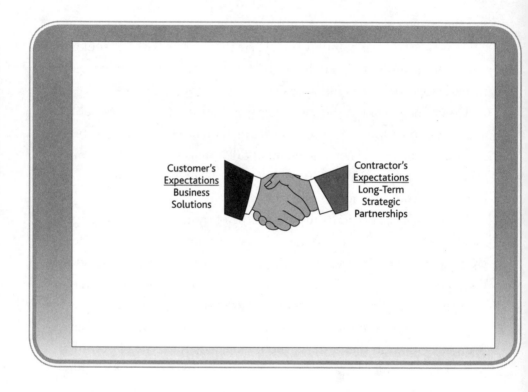

Complex projects have been in existence for decades. But only recently has the term *engagement project management* or *engagement selling* been used, especially for complex projects. In the past, the salesperson would sell products or services to a potential client, and after the sale is completed, the salesperson would move on to the next client. Today, salespeople are encouraged to maintain relationships with clients to see what other products or services can be provided.

In a courtship that leads to marriage, an engagement can be viewed as the beginning of a lifelong relationship. The same holds true for engagement selling. Customers are undertaking more complex projects each year and must rely heavily on contractors for support. What customers want is someone to provide them with solutions to their business problems.

Contractors, however, are willing to develop superior project management capability to become a solution provider and want to remain a strategic partner with the customer forever. In other words, the customer wants a solution provider, and the solution provider wants a long-term partnership arrangement. Just like in marriage, finding the right partner can satisfy each one's needs.

Solution providers are usually willing to custom-design their project management systems, forms, guidelines, templates, and checklists for a particular client in hopes of a long-term strategic relationship. The solution provider can assist the customer in the strategic planning activities for the next complex project.

BEFORE AND AFTER ENGAGEMENT PROJECT MANAGEMENT

Before Engagement Project Management	After Engagement Project Management
Continuous competitive bidding	Sole-source or single-source contracting (fewer suppliers to deal with)
Focus on near-term value of the deliverable	Focus on lifetime value of the deliverable
Contractor provides minimal support for client with their customers	Support client with their customer value analyses (CVAs) and customer value measurements (CVMs)
Utilize one inflexible, linear EPM system	Access to contractor's many nonlinear systems

As mentioned previously, successful engagement project management can be a win-win position for both parties. Perhaps the greatest advantage to engagement project management is that the cost of competitive bidding is minimized. The solution provider is treated as a single-source or sole-source provider and, because of the strategic partnership, may not be required to submit formal proposals for the next complex project.

Unlike traditional project management, where the prime concern is the deliverable handed to the customer in the short term, solution providers have a long-term relationship with the client and are interested in the long-term value of the complex project's deliverables. Long-term rather than short-term support is provided. In addition, new forms of strategic relationships, such as build-operate-transfer (BOT) and collaborative working arrangements (CWAs) are emerging to meet these needs. Undoubtedly, additional forms of these arrangements will develop over time.[2]

Previously, we mentioned the importance of value as a driver for success. Customers are now implementing value analysis and value measurement programs not only internally, but also to support their own customers. In this regard, the business solution provider can provide support to the customer for a variety of value programs.

---

[2] See Chapter 7 for further discussions of various strategic alliances (e.g., BOT, CWA).

PERCENTAGE OF PROJECTS USING PROJECT MANAGEMENT

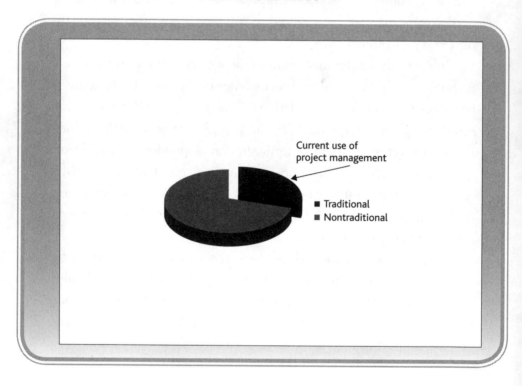

For more three decades, project management has been used to support traditional projects. Previously, we discussed the difference between traditional and nontraditional or complex projects. Traditional projects are heavily based on linear thinking; we have well-structured life-cycle phases and templates, forms, guidelines, and checklists for each phase. As long as the scope is reasonably well defined, traditional project management works well.

Unfortunately, only a small percentage of all of the projects within a company fall into this category. The larger percentage of nontraditional or complex projects use seat-of-the-pants management because they are largely based on business scenarios where the outcome or expectations can change from day to day. As such, project management techniques were neither required nor used on these complex projects that were more business oriented and aligned to 5-year or 10-year strategic plans that were constantly updated.

Now, we are finally realizing that project management can be used on these complex projects, but the traditional project management processes or techniques may be inappropriate or must be modified. The leadership style for complex projects may not be the same as with traditional projects. Risk management is significantly more difficult on complex projects, and the involvement of more participants and stakeholders is necessary.

POSSIBLE COMPLEX PROJECT OUTCOMES

Ability to Manage Complex Projects		
Client	Seller	Possible Results
Good	Good	This is the best of both worlds. The projects have a high probability of success through a strong partnership.
Good	Poor	Client must provide close governance. Seller may not have an appropriate EPM system.
Poor	Good	This can lead to possibly disastrous consequences if the client tries to micromanage the seller.
Poor	Poor	Chances of success are very limited. Significant cost overruns will occur.

I s it possible that both the customer and the solution provider have the same degree of knowledge about how to manage complex projects? It is possible but highly unlikely. In the preceding feature, you can see the four possible scenarios related to the buyer's and seller's knowledge. Although four scenarios are possible, it is unlikely that a customer would ever select a contractor or solution provider that did not possess the capability and experience to manage complex projects. The exception is when a customer in a developing nation is mandated by politics to use a local firm to take the lead as the project manager. While less frequent, this problem can occur in an industrialized nation as well.

Not all customers have expertise in managing complex projects. If the customer did have such expertise, then why would the customer need the contractor? Customers must trust solution provider contractors to do the job effectively. Without such trust, the chances of success are diminished.

The greater the complexity of a project, the greater the number of complex systems that must interact. If mistrust prevails, the interactions among the complex systems can be prolonged. Complex projects must be managed in an atmosphere of trust. This is particularly important if virtual teams are needed.

LONG-TERM GLOBALIZATION PROJECT MANAGEMENT STRATEGY

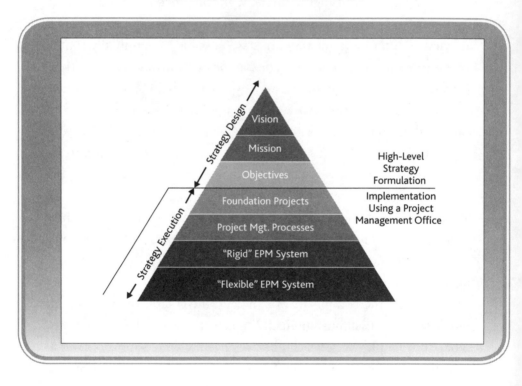

It is important to understand what types of companies function as solution providers for complex projects. In the preceding feature, you can see the evolution of project management in a solution provider. The importance of project management must be recognized at the executive levels. The result is a vision, mission, and objectives for project management implementation.

In the next step, the foundational or traditional projects in which project management will be used are identified. These may be small or breakthrough projects where people can actually see project management in action.

The next step is the development of project management processes to support each phase of the project. Once sufficient processes are created, they are combined into a rigid EPM methodology to support the management of traditional projects. It is at this point where we separate the average company with the solution providers.

Solution providers understand that one methodology will not satisfy all clients, especially if a long-term partnership is desired. Solution providers must employ highly flexible methodologies that can be adapted and custom-designed to a particular client. There can be a distinct methodology for each client. As best practices and lessons learned are obtained from the completion of a complex project, processes are undated through a continuous improvement practice. The flexible methodology can have key performance indicators (KPIs) that are unique to that customer or that particular complex project. The closer that the methodology is aligned to the customer's business processes, the more likely it is that the customer will see the value in the strategic partnership relationship.

GLOBAL VERSUS NONGLOBAL COMPANIES

Factor	Nonglobal	Global
Core business	Sell products and services	Sell business solutions (value)
Project management satisfaction level	Must be good at project management	Must excel at project management
Project management methodology	Rigid	A framework with flexibility
Type of team	Co-located	Virtual
Supporting tools	Minimal	Extensive
Continuous improvement	Follow the leader	A necessity for survival

Every country in the world has complex projects, but not every country has resources qualified to manage these complex projects. Therefore, those companies that have taken the time and effort to develop flexible project management methodologies and become solution providers are companies that are competing in the global marketplace. Although these companies may have products and services that they can provide as part of their core business, they view their future as being a global solution provider for the management of complex projects.

For these companies, being good at project management is not enough; they must excel at project management. They must be innovative in their processes to the point that all processes and methodologies are highly fluid. They have an extensive library of tools to support the project management processes. Most of the tools were created internally with ideas discovered through the capturing of lessons learned and best practices. They have a robust improvement and innovation process in place—where lessons learned are shared among the entire organization in a planned, effective manner. They must make a significant investment in knowledge management to facilitate the instantaneous collection of these lessons learned data and their equally rapid distribution among all project teams.

QUANTITY OF TOOLS

Quantity of Tools

Areas Where Multinational Companies Are
Developing Supporting Tools for Managing
Complex Projects

Level #3: Contracting,
Overall Business Mgt.,
and H.R. Management

Level #2: Risks, Decision-
Making Techniques, and
Communication Mgt.

Level #1: Planning
(Scope Management),
Cost and Schedule

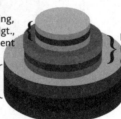

In the previous feature, we stated that solution providers for complex projects have a vast library of tools to support each and every complex project. The preceding feature shows the three levels of tools that are being developed for the management of complex projects.

Level 1 is the traditional level that almost all companies possess. These are the tools for the planning, scheduling, and cost control of projects. What differentiates the solution providers of complex projects is that they use project dashboards with real-time status reporting. This is highly advantageous to the customer. Cost and schedule reporting is done through report generators where the information will be presented in a form desired by the customer.

Level 2 contains tools appropriate for communicating with virtual teams, risk management, and decision making. Once again, these tools are custom-designed for the particular risks with which a customer must deal and the types of decisions appropriate to that customer.

Level 3 tools can be both specific and general. Contracting tools may be generic. However, it is not uncommon for a solution provider to create business-related tools to support the customer, especially if the solution provider sees a long-term relationship with the customer.

PROJECT MANAGEMENT SOFTWARE

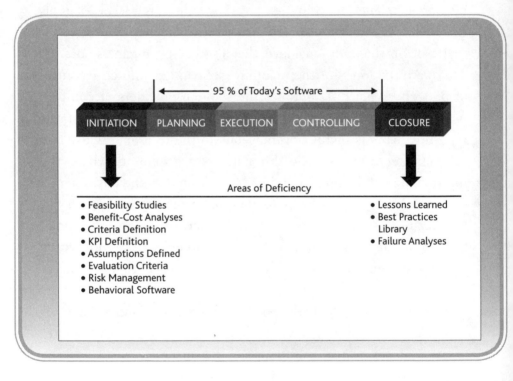

The preceding feature shows the breakdown of project management software. Today, perhaps as much as 95 percent of the project management software is designed around the planning, scheduling, and cost control of projects. While much of this software is also applicable to complex projects, it is the other 5 percent that may be crucial for complex projects.

There is a shortage of project management software for project initiation and project closure. For complex projects, project initiation may very well be the most important life-cycle phase. Solution providers for complex project are creating their own software for the initiation phase and the closure phase. They are also creating behavioral software for the evaluation of people that may be well suited to work on virtual project teams.

In the past few years, integrated project management tool sets have been appearing on the market. Instead of a project management team having to figure out how to integrate the data and methods of several tools from different software development manufacturers, these new tools provide for integrated data and methods, thereby facilitating the collection, analysis, distribution, and reporting of project management indicators among team members, management, and other key stakeholders in a relatively secure web-based environment (if such an environment indeed exists).

AREAS OF BEST PRACTICES

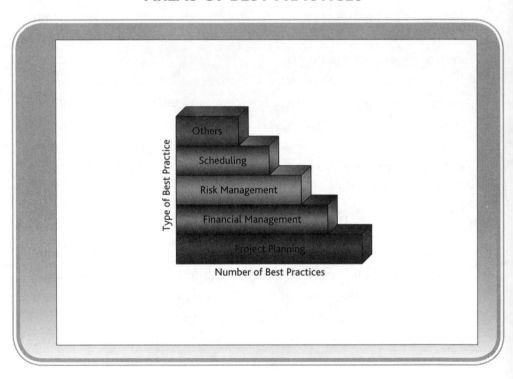

Complex projects undergo debriefings for best practices the same way traditional projects do. The difference is that the best practices discovered may be applicable just to this type of complex project and/or client rather than being generic in nature.

As seen in the preceding features, most of the captured best practices seem to be related to the planning activities. The best practices are most likely related to the forms, guidelines, templates, and checklists associated with the planning processes.

The debriefing of complex projects can be heavily focused on financial management and risk management. This is to be expected. The debriefing session will most likely address the identification of all of the risks that impacted this complex project. It is unlikely that all of these risks were known at project initiation.

Furthermore, on complex projects, it may no longer be acceptable to wait until the end of a phase or for project closing to collect best practices. As mentioned previously, all of these activities associated with project practices (planning, risk management, etc.) are performed iteratively throughout the project. And because of the extended duration of many complex projects, it's imperative that these emerging best practices be captured as soon as they are identified, codified, and rapidly distributed to the appropriate project stakeholders so they can be implemented as quickly as possible on the current project.

THE COLLECTIVE BELIEF

- Inability or refusal to recognize failure
- Refusing to see the early warning signs of possible disaster
- Seeing only what you want to see
- Fearful of exposing mistakes
- Viewing bad news as a personal failure
- Viewing failure as a sign of weakness
- Viewing failure as damage to one's career
- Viewing failure as damage to one's reputation

Long-term, highly complex projects often mandate that a collective belief exists. The collective belief is a fervent, and perhaps blind, desire to achieve that can permeate the entire team, the project sponsor, the stakeholders, and the most senior levels of management. The collective belief can make a rational organization act irrationally by refusing to hear bad news, refusing to be willing to cancel a project, and other such faulty arguments.

When a collective belief exists, people are often selected for the complex project teams based on their willingness to support the collective belief. People are not allowed to challenge results, and bad news is often hidden. As the collective belief grows, nonbelievers are trampled and eventually forced off of the project.

The collective belief often makes it difficult to cancel projects. However, there are other reasons why some projects are difficult to cancel. These items are shown in the preceding feature. Not all complex projects will be successful. Some must be canceled, and the earlier they are canceled, the quicker resources can be assigned to projects that offer a greater opportunity for organizational success.

Chapter

2

INTEGRATION MANAGEMENT

CHANGES IN FOCUS

Managing Traditional Projects	Managing Nontraditional Projects
Focused on project controls and balancing the triple constraints	Focused on excellence in leadership, motivation, and communication

On a traditional project, where we may have one and only one stakeholder, and that stakeholder is the customer or the project sponsor from the funding organization, the focus is on project controls and the triple constraint. We do admit that we might not always be able to accomplish the project within the triple constraints, but we are willing to balance the constraints as best we can.

With nontraditional or complex projects, where we may have a large number of stakeholders, the emphasis is on excellence in communications, leadership, and motivation. On traditional projects, there is a high expectation that we can achieve the triple constraints. And even if we cannot get the job accomplished within the triple constraints, we probably will still get follow-on work from this client. On large, complex projects, there is an expectation of schedule slippages and cost overruns. The larger the project, the greater the cost overrun and schedule slippage. An inability to meet the triple constraint on large projects will not prevent the contractors from receiving follow-on work. But failures in leadership, motivation, and especially stakeholder communications will almost assuredly guarantee no follow-on work. Not all stakeholders understand the triple constraints or the project management tools, but they do understand communications, leadership, and motivation.

PROJECT SPONSORSHIP (1 OF 2)

Managing Traditional Projects	Managing Nontraditional Projects
Project sponsorship may be a single individual most likely from the funding organization and may or may not reside at the executive levels.	Project sponsorship will be replaced by complex governance and fragmented throughout all of the stakeholders and most likely will include executive management and possibly senior government officials.

In traditional organizations, sponsorship was provided by the funding organization and the sponsor usually resided at the executive levels of management. The role of the sponsor was to oversee all decisions made by the project manager. As project management began to mature, executives began having more trust in the project manager. Sponsorship on some projects was now at the middle levels of management rather than at the senior-most levels. Also, as the organization undertook more projects, senior management recognized that they could not act as sponsors on all projects and still carry out their normal duties.

Some project managers preferred having middle management as sponsors rather than senior management. Middle management was more readily accessible than senior management, and middle management had a better understanding of the technology such that faster decisions could be made.

On complex projects with many stakeholders, each stakeholder may view themselves as a sponsor. All of the stakeholders, whether active or passive stakeholders, make up the project's governance group. On complex projects, sponsors and stakeholders may not be in agreement on the solution to a problem. This is particularly true on large projects with large governance groups.

PROJECT SPONSORSHIP (2 OF 2)

Managing Traditional Projects	Managing Nontraditional Projects
Decision-making techniques such as facilitative workshops, group creativity techniques, and alternative identification are all possible and can be accomplished in a timely manner.	Because of the number of possible stakeholders and the limited information that each one may possess, making decisions in a timely manner may not be possible regardless of the techniques used.

The size of the stakeholder community on a project determines the decision-making techniques that will be used. With traditional projects, where we have a limited number of stakeholders, techniques such as facilitative workshops, group creativity, and alternative identification are all possible and can be done in a timely manner.

On nontraditional projects with a large number of stakeholders, and where each stakeholder may possess limited knowledge concerning the issues and problems, getting agreement may be difficult. Using groupthink processes may not work, either, because there usually exists a dominant voice in the group and then there is a tendency for those who have no opinion to side with whomever they believe will win the argument even though they may not believe in what the individual is advocating.

One of the better solutions is to work with the key or influential stakeholders, assuming there are not too many of them. This is why it is essential to identify the key stakeholders, or else the decision-making process may become elongated.

PROJECT ACCOUNTABILITY

Managing Traditional Projects	Managing Nontraditional Projects
Accountability generally resides with the project manager, but in some cases can be shared with the line managers.	Shared accountability may permeate the entire project. Likewise, shared leadership may permeate the entire project even if single-person accountability exists.

On traditional projects, especially smaller projects, accountability resides with the project manager. However, that trend appears to be changing in companies where the project manager possesses an understanding of technology rather than a command of technology. Project team members who have technical issues generally migrate to the person with the technical expertise, typically the line manager. Simply stated, when line managers are providing daily direction to their workers, the line managers are held accountable for the deliverables they must provide to the project. Therefore, line managers and the project manager now share accountability for the success and failure of the project. This will work if the concept of shared accountability is enforced by senior management. This is one of the reasons why today line managers are becoming Project Management Professionals (PMPs).

On complex projects, accountability is almost always shared between the stakeholders and the project team. Regardless of what we read in textbooks, single-person total accountability on large, complex projects may not be possible. The problem is further complicated when a large portion of the project is managed by virtual teams that may be under the control of the stakeholders rather than the project manager.

On some complex projects, particularly those involving leading-edge technologies, there may be some version of what's called the "leaderless team" approach. While the project manager still has the ultimate accountability for the project, project leadership shifts among the key project management team players, depending on which phase of the project is being executed. For instance, if the project used a sequential life cycle (although the leaderless team concept isn't limited to this type of life cycle), the design phase might be led by a senior design engineer, the development of the product by a senior manufacturing engineer, and so on.

EPM METHODOLOGIES

Managing Traditional Projects	Managing Nontraditional Projects
Perhaps as little as one standard, rigid methodology exists for all projects.	There may be a necessity for the development of multiple EPM methodologies based on the complexity of the project at hand.

Companies have come to the realization that one of the critical factors for project management success is a good enterprise project management (EPM) methodology.

Some companies are fortunate in that they can use one EPM methodology for all projects, whereas other companies may require more than one. An auto manufacturer may use multiple methodologies—for example, one for new product development and another for information technology (IT)-related projects.

Singular methodologies provide a structured process for the management of projects. Life-cycle phases are created, and each phase is accompanied by forms, guidelines, templates, and checklists. This structured process, or linear thinking, works well if the projects all have some degree of similarity.

For complex projects with multiple stakeholders, process similarities may be nonexistent. The contractor may find it necessary to have multiple methodologies or to go so far as to create a separate process for a client in hopes of acquiring follow-on work. The concept of customized processes on complex projects is growing in acceptance. Each customized methodology may require special tools, dashboards, and key performance indicators (KPIs). Companies that provide business solutions have a large collection of tools that project managers can use.

ENTERPRISE ENVIRONMENTAL FACTORS

Managing Traditional Projects	Managing Nontraditional Projects
Because the project's duration is usually 18 months or less, it is reasonable to assume that no major changes will occur in the enterprise environmental factors. The environment may be regarded as known.	The enterprise environmental factors will change over the life of the project. The changes may affect each stakeholder differently and create problems for the project manager. The environment may be volatile and chaotic.

According to the *PMBOK® Guide*, enterprise environmental factors are any or all of the environmental factors that can affect the organization in its execution of the project or the way in which it views project success. The factors include culture, structure, governance, market conditions, available resources and even available software. For traditional projects that are usually less than 18 months or so, the enterprise environmental factors can impact the project, especially if they change over the duration of the project. The relative change is dependent on the duration of the project, all other things being equal.

On complex projects with long time frames, the enterprise environmental factors must be continuously monitored. They can and will change over the duration of the project. These factors can change in each stakeholder's organization and country. It is essential that each stakeholder monitor these in their community and provide feedback to the project manager. It is impossible for the project manager to possess the capability to monitor these factors everywhere. Changes in enterprise environmental factors can convert key stakeholders into inactive stakeholders or observers, and vice versa.

As value-driven project management becomes more salient, organizational goals and market conditions, which are bound to change over a prolonged period of time, become more important enterprise environmental factors. If a company wants to ensure that it obtains value from the project's end product, it needs to monitor both of these to ensure that the product will meet its original (or changing) long-term market value and continues to accord with stated organizational goals. If it does not, project termination should be considered.

ORGANIZATIONAL PROCESS ASSETS

Managing Traditional Projects	Managing Nontraditional Projects
The project team is reasonably knowledgeable on the available organizational process assets.	Not all team members will have the same organizational process assets. Each team member's company may be at a different level of maturity in project management. For reporting purposes, the host may not be able to interface with the project team's organizational process assets.

Organizational project assets are the tools available to the project manager. The tools may be corporate policies, procedures, guidelines, forms, templates, and checklists, as well as the EPM system itself. The project manager may also develop project-specific tools, but it is unlikely this will happen on the traditional-type projects. On traditional projects, all of the tools usually reside in the contractor's organization.

On complex projects, each stakeholder may be at a different level of project management maturity and therefore possess a variety of different project management tools, many of which may be obsolete or simply not be capable of interfacing with the project manager's tool box. Based on the length of the project, and the fact that the contractor wants follow-on work from these stakeholders and clients, the project manager may be willing to share his/her tools as well as developing organizational process assets for the stakeholders.

It is not uncommon for project managers to train clients and stakeholders on the project manager's organizational process assets and their use. If real-time dashboards are used, the project manager may provide training on their use and data interpretation.

In some cases, the owner may dictate to some extent the types of tools that all contractors are required to use for progress reporting and other processes (e.g., risk management tools) and for the specific types of output they expect to see from these tools (e.g., Gantt charts, precedence diagrams, Monte Carlo simulation curves, tornado diagrams).

WEAKNESSES IN LEADERSHIP SKILLS

Managing Traditional Projects	Managing Nontraditional Projects
Weaknesses in the leadership style of the project manager can be compensated for with a strong sponsor and good line managers that may share the leadership responsibility.	Weaknesses in the leadership of the project manager can bring the project to its knees.

In traditional projects, the project manager resides in the parent company and may be closely supervised and mentored by the project sponsor. Weaknesses in project management leadership can be compensated for by the sponsor or others. Also, project managers with subpar leadership capability can be replaced with little damage to the ongoing project.

On complex projects that may be multinational and where the project manager is physically removed from the parent organization, poor leadership can be costly. Each stakeholder can have their own opinion of what effective project management leadership means, and their views may be contradictory to the actual leadership style used by the project manager. Project managers who have good leadership styles for traditional projects may discover that the same leadership style is ineffective on complex projects.

Most traditional project managers may have one leadership style that they feel comfortable using, especially if it is accepted by the sponsor or client. But on complex projects, the project manager may need a different leadership style for each stakeholder because the leadership style must match the expectations of the stakeholders, especially the key or influential stakeholders. Also, the project manager's leadership style may need to change based on the life-cycle phase.

PROJECT'S BUSINESS CASE

Managing Traditional Projects	Managing Nontraditional Projects
The business case is prepared by the user and the project manager, and reviewed by the project management office (PMO) for alignment to the corporate objectives.	Because of the number of stakeholders, business case development may be complex due to the input requirements from a multitude of stakeholders. The project manager may not be included in the development process, and the PMO may not be involved in the review.

In companies that focus on traditional projects, the business case for the project may be reviewed by the project management office (PMO) to make sure it is aligned with corporate objectives, and the project manager may even be involved in preparation of the business case to make sure that the timing, funding, and overall expectations are realistic. There is a tendency today for project managers to be brought on board relatively early during the initiation phase because project managers today possess more business knowledge than did project managers of years ago.

On complex projects, business case development may be a long, tedious process involving many stakeholders. Many stakeholders may succumb to accepting a poor business case and later push for changes to be made. Not all stakeholders are equal when it comes to preparing a business case on a complex project. There is also the possibility that certain key stakeholders are not involved in the preparation of the business case but should be.

As noted in other sections, the prolonged duration of complex projects may require regular evaluations of the original business case in the light of current business conditions. Business conditions at the beginning of the 21st century change quickly. Projects with durations of 5 to 10 years may see a number of changes in those conditions, changes in market demand for types of products, and changes in technologies able to produce those products. So it's vital for the project portfolio management team to regularly review the probable value of the project's end product in view of these changing conditions.

PROJECT GOVERNANCE

Managing Traditional Projects	Managing Nontraditional Projects
Project governance is well understood and will usually remain the same for the project's duration.	There is a risk that the governance will change. This can create obstacles to projects with respect to decision making and meeting objectives. There must be a sustainability of governance.

Project governance is designed to provide a consistent and comprehensive method for controlling the project to maximize the probability of its success.[1] It consists of business rules and processes that apply from the organizational to the project level. On traditional projects, mainly because of the relatively short time duration, project governance can be expected to remain the same for the duration of the project. On long-term complex projects, governance will change over the duration of the project. The longer the project, the greater is the possibility that those who are stakeholders at the beginning of the project will not be the same stakeholders as those at the end of the project. And as we stated previously, stakeholders can change during the project because of political changes in their country, promotions, resignations, and retirements.

Governance can also change when a project gets into trouble. Some stakeholders may wish to be more actively involved in the governance process, whereas other may act like rats deserting a sinking ship.

Stakeholder management is a difficult task for project managers, even on traditional projects. It becomes much more of a challenge on complex projects. The process of conducting a stakeholder analysis leading to the development of a stakeholder register is both more difficult and more important on complex projects. Once developed, this register has to be reviewed at regular intervals throughout the project due to the stakeholder changes for reasons noted earlier.

---

[1]Adapted from Project Management Institute, *A Guide to the Project Management Body of Knowledge*, 4th ed. Newtown Square, PA: Project Management Institute, 2008, p. 20.

PROJECT'S ASSUMPTIONS

Managing Traditional Projects	Managing Nontraditional Projects
The project's assumptions are reasonably clear and documented in the project charter. The assumptions are well understood and revalidated throughout the project.	The project's assumptions are not always documented and are often taken for granted. Revalidation will occur, if at all, during rejustification of the project.

On traditional projects with a reasonably short time duration, the assumptions can be expected to be clear and well understood. The assumptions may undergo a revalidation process, but it is unlikely that they will change. If changes are necessary, the impact on the project may be minimal.

On complex projects, the assumptions will change and can have a major impact on the project. Changing assumptions may result in the cancellation of the project. The project plan may include provisions for periodic reevaluation of the project's assumptions.

The assumptions that are made may impact each of the stakeholders differently. Therefore, stakeholder involvement in revalidation of assumptions is critical. This includes the key stakeholder as well as the observers that may be affected.

In many projects, careful examination of project assumptions and the uncertainty surrounding them is a basis for developing a list of known project risks. The prolonged durations of complex projects require the project management team's vigilance in monitoring the risks associated with assumptions that are developed by project team member and stakeholders throughout the project.

ALIGNMENT OF GOALS

Managing Traditional Projects	Managing Nontraditional Projects
Corporate goals and project goals are the same.	Corporate goals and project goals are not aligned. Each stakeholder group can have a different goal in mind for the project, thus creating havoc with decision making.

Project goals and corporate goals may not be aligned on both traditional and complex projects. However, the misalignment is more prevalent on complex projects because of the large number of stakeholders, each of whom may have their own agenda. The fact that larger, more complex projects may be in the spotlight doesn't mean that the project's goals are aligned with strategic objectives.

Complex projects have a greater tendency to be influenced by political decisions of certain key stakeholders. This can place the project manager in a precarious position when critical decisions must be made.

Additionally, the extended durations of complex projects means that these projects will be executed through times during which corporate goals will change, resulting in the misalignment noted above. Since these goals inform the prioritization among the project's competing demands (scope, schedule, cost, quality, risk, resources), the project manager may occasionally need to seek guidance from the project sponsor and other key stakeholders to minimize the disparity between project and corporate goals and to keep the project heading in the right direction through appropriate trade-offs.

EXPERT JUDGMENT

Managing Traditional Projects	Managing Nontraditional Projects
With traditional projects, there exists a small group of subject matter experts to support the expert judgment needs of the project.	There may be teams of subject matter experts affiliated with each of the stakeholder groups. Getting all of the subject matter experts to come to any agreement may be complicated.

With traditional projects, generally there will exist only a small group of subject matter experts to support the project. The small group can be from within the company or hired through contractors. Most important is that the quality of the subject matter experts is known.

On complex projects, each stakeholder group can have their own group of subject matter experts, but not all subject matter experts are the same, even within the same technical discipline. It is possible that a senior engineer in a developing nation would be equal to a junior engineer in an industrialized nation.

With a multitude of subject matter experts, it may be very difficult to get agreement on the solution to a problem. Some subject matter expert groups may be heavily biased by their stakeholders or superiors to promote a solution that is not in the best interest of the entire project. Politics, culture, religion, and perceived status can influence the decisions of these groups.

PROJECT CHARTER

Managing Traditional Projects	Managing Nontraditional Projects
The company may have a template for the project charter. The project sponsor signs the project charter and there is agreement on what is contained in the charter.	The authority needed for managing the project will most likely be diluted over several project managers. Each stakeholder may wish to control the authority of the project manager in their group. There may be no agreement as to what information is contained in each of the charters.

On traditional projects, there may exist a template for the preparation of the charter. The template identifies what should be included in the charter but may include other items that are company specific. The charter is signed by the project sponsor, but it may be prepared by the project manager.

On complex projects, there may be multiple charters, one for each stakeholder's organization and possibly each virtual project team. Each charter can contain different information. Although the charter supposedly identifies the authority of the project manager, each key stakeholder may wish to control the authority of the project managers under their control. The result may be that certain parts of the project team have very limited authority, while other parts of the team may have a tremendous amount of authority.

PROJECT DECISION MAKING

Managing Traditional Projects	Managing Nontraditional Projects
The project charter may specify the authority of the project manager with regard to decision making. Usually, the project manager and the project sponsor control the decisions.	The project manager may have limited authority with regard to decision making. The greater the number of stakeholders, the more likely it is that consensus decision making will take place.

Project charters generally identify the authority of the project manager but do not necessarily identify the decisions that the project manager can make. On traditional projects, the project manager and the project sponsor work closely together, and decision making may be a joint effort.

On complex projects, the decision-making process can be long and drawn out because of the vast number of stakeholders. In some countries, even the key stakeholders may not have the authority to make any decisions. The more complex the project, the greater the tendency that the decisions must be made or at least approved at the highest levels of management, perhaps even in the government.

Even though we call certain stakeholders influential stakeholders, it by no means implies that they have the authority to make decisions. They may appear as influential but they actually monitor performance and report it to a higher level for decision making.

The greater the number of stakeholders, the fewer the options available for group decision making. It may seem appropriate to place all of the stakeholders in a room to arrive at a decision, but that assumes that all of the stakeholders are empowered by their superiors to make decisions. Also, the authority that the stakeholders possess can be based on the type of decision to be made. Some stakeholders may be authorized to make technical or scope decisions but not financial decisions.

GO AND NO-GO DECISION POINTS

Managing Traditional Projects	Managing Nontraditional Projects
Only a small percentage of the traditional projects get canceled prior to the original completion date.	Because of the amount at stake, there must exist "off ramps" throughout the project where the project can be terminated.

The majority of traditional projects never get canceled. This is because the assumptions are reasonably well known and the enterprise environmental factors are relatively constant over the life of the project. The project may be completed late and over budget, but it will be completed.

Complex projects generally have large cost overruns and schedule slippages. The larger the project, the larger the cost overrun and schedule slippage. It is important on large projects to use life-cycle phases because the end of each phase is an "off ramp" should the project need to be canceled.

The most difficult decision facing an executive is the cancellation of a project because it could prove to be an embarrassment for the executives and stakeholders that avidly promoted the project. Large projects where the objectives cannot be met should be canceled as soon as possible to limit the damage. There are tell-tale signs that a project is in trouble or is about to be canceled such as the unannounced reassignment of critical stakeholders and/or replacing them with lower ranking individuals. As a corollary, replacing a junior person with a senior person as a stakeholder is usually a sign of continuation of the project.

PROJECT REPLANNING

Managing Traditional Projects	Managing Nontraditional Projects
Project managers spend a great deal of time planning the project and try to maintain the same plan throughout the project. If replanning does take place, it is with the intent of compressing the schedule.	Project replanning is a way of life. Not all stakeholders are skilled in planning their own portions of the work. Replanning is caused by a change in stakeholders, unclear goals, and a change in the political climate. Project plans will evolve throughout the project.

Once the project is planned and execution begins, project managers on traditional projects spend a great deal of their time looking for ways to replan the project. Replanning generally focuses first on schedule compression and second on cost reduction. This can occur throughout the life cycle of the traditional project.

On complex projects, project managers are at the mercy of the stakeholders for assistance in replanning activities. Once again, not all stakeholders are equal in their ability to perform replanning activities. With complex projects, replanning activities may include a large number of stakeholders, all of whom must agree on the changes and the expectations. This may be difficult, if not impossible, since all of the stakeholders may have different goals and objectives.

Replanning is sometimes caused unexpectedly because of politics. In one instance, a politician was up for reelection and demanded that a large construction project begin in his district so that he could use the project as a campaign weapon to attract more votes. This caused funds to be diverted from other projects, thus forcing other projects to be either placed on the back burners or descoped because of cost reductions.

OPTIMISM

Managing Traditional Projects	Managing Nontraditional Projects
Optimism, even excessive optimism, can exist but is usually corrected quickly because of the reasonably short duration of the projects.	Excessive optimism can lead to significant cost overruns. Excessive optimism leads to withholding of adverse information and a fervent desire to let the project continue even though it should be canceled.

Most project objectives as well as project plans are based on optimism. On traditional projects, if the optimism is overstated and needs to be corrected, it can usually be done quickly. But on complex projects, quick fixes to optimism may be impossible.

The larger the project, the greater the optimism. Long-term, highly complex projects often mandate that a collective belief exists, as was discussed in the framework section of this book. The collective belief is a fervent and perhaps blind desire to achieve and meet the optimistic objectives. This optimism is often a necessity, and the entire team—the project sponsor, the stakeholders, and the most senior levels of management—must have this optimism and fervent belief to pursue the project under difficult circumstances. Unfortunately, the greater the optimism, the greater the tendency that a rational organization will act irrationally by refusing to hear bad news, refusing to be willing to cancel a project, and other such faulty arguments.

When a collective belief exists, especially if optimism is high, people are often selected for the complex project teams based on their willingness to support the collective belief, and the people must have the same level of optimism.

The greater the optimism, the more difficult it is to cancel projects. However, there are other reasons why some projects are difficult to cancel.

POOR PROJECT PERFORMANCE

Managing Traditional Projects	Managing Nontraditional Projects
With poor project performance, the project manager is usually given the flexibility to develop contingency plans in order to salvage the project.	With poor performance, the project must be rejustified before any attempt is made at contingency planning. This will create a lengthening of the project.

Not all projects are completed according to the project plan. Some clients may be willing to accept a project's deliverables, even though not all of the specifications were met. This is particularly true on traditional projects.

On complex projects that are suffering from poor performance, rejustification must take place periodically. Contingency planning is mandatory and an essential component of the project management plan. Not all projects will have great performance. On complex projects with a lot of money at stake, poor performance forces stakeholders to consider project cancellation rather than throwing good money after bad money.

Poor performance that is recoverable is good. Poor performance that is not recoverable is bad, and this is when stakeholders should consider limiting their losses. It is important to ascertain the root cause of the poor performance and its potential impact on the morale of the team.

PROJECT JUSTIFICATION

Managing Traditional Projects	Managing Nontraditional Projects
Once approved, projects continue on unless performance mandates that the project be canceled.	Projects exist just for the duration of the funding cycle, which may be yearly. The project may then have to be rejustified, especially if politics or a new administration appears.

Project justification should be based on valid objectives that are aligned to the business or strategic goals of the stakeholders. On traditional projects, once the project is justified, it continues on, usually without any rejustification, unless serious problems emerge that can mandate the project be canceled.

Large, complex projects exist for the duration of the yearly funding cycle. At the end of each year, project rejustification is necessary for the next yearly funding cycle. Even if the rejustification is completed successfully, politics may dictate that the funds be redirected to some other project. This can happen even if funding was allocated for the entire project.

Some of the critical factors that mandate periodic rejustification are changes in key stakeholders, changes in the political administration of the host country, changes in critical project assumptions, and changes in the enterprise environmental factors discussed previously. Additionally, the value to the organization of the project's outcome may change due to changing business or market conditions. In organizations that have a robust project portfolio process in place, all projects will be normally be reviewed on a regular basis in the light of the current market conditions and enterprise goals. Those projects that do not continue to meet these changing value requirements should be canceled.

PROJECT PLAN OWNERSHIP

Managing Traditional Projects	Managing Nontraditional Projects
The project team creates the plan and owns the plan.	The plan is most likely created by various stakeholder groups, and the team views their role as the execution of a plan created elsewhere. Motivation may diminish because of lack of ownership.

On traditional projects, the project team creates the plan and maintains the ownership for the plan. However, on complex projects, the plan may very well be developed by multiple stakeholders, and the project manager may simply function as a coordinator or integrator of plans.

People who develop plans generally feel ownership and demonstrate a strong loyalty to the plan. If people do not feel ownership, motivation may diminish. Having stakeholders sign off on the plan may help, but the risk still remains. There may, however, be other ways to mitigate some of this risk.

Regardless of who develops the plan, if the plan is reviewed and discussed in detail with the various project team members, their concerns and disagreements with specific parts of the plan can be reviewed. The project manager can then discuss those areas of the plan with the originating stakeholders and push to have the appropriate changes made. In order to accomplish this successfully, the project manager must understand the implications of the plan in its current state, the team members' concerns, and the impact on the competing demands of making the requested changes. Only by doing so can the project manager then have constructive discussions with the relevant stakeholders.

THE PROJECT PLAN: SUMMARY LEVELS

Managing Traditional Projects	Managing Nontraditional Projects
The project plan is usually well defined and undergoes minimal changes throughout the life of the project.	The project plan, if it exists at all, may be ill defined and may be periodically changed based on politics, elections, reductions in funding, and changes in stakeholders.

On traditional projects, the summary or macro-level plans are usually well defined and may not change over the life cycle of the project. If scope changes occur, the effects may be at the detail levels rather than at the summary levels.

On complex projects, summary-level plans, especially major milestones, may be established by key stakeholders before the detail plans are developed. The result is that the predetermined milestones may not be realistic, and yet key stakeholders may force the project manager to accept these milestones because of factors such as politics, upcoming elections, funding constraints, and other such arguments.

Summary-level plans are more likely to be impacted by scope changes on complex projects. The more complex and costly the project, the greater the impact of scope changes on all levels of the plan. Also, changes in the enterprise environmental factors can lead to changes in the summary-level plans.

Project managers are usually more actively involved in the detail planning than in the summary-level planning. The summary-level milestones may be predetermined in the request for proposal (RFP) and proposal statement of work and set forth as a requirement in the bidding process, even though detail plans cannot be established to meet these milestones. Good stakeholder management can minimize the impact of this situation.

PROJECT MANAGEMENT PLAN

Managing Traditional Projects	Managing Nontraditional Projects
The project manager, working with the functional leads or assistant project managers, will prepare the project management plan.	Development of a single project management plan may be impossible because of the complex agreements needed. The plan may be accomplished using rolling wave planning.

On traditional projects, the project manager works with the assistant project managers and the functional leads during the initial project kickoff meeting to develop the project management plan. Several meetings may be required to do this based on the worker skill levels needed, availability of resources, and technology requirements. The result is normally a single project management plan.

On large, complex projects, it may not be possible to develop a single project management plan that covers the duration of the project. The project management plan is usually broken down into subsidiary plans such as a facility utilization plan, cost plan, quality plan, procurement plan, testing plan, resource management plan, staffing plan, and other such plans.

Stakeholder involvement may be essential in the development of the subsidiary plans. Getting stakeholder agreements may be difficult. Furthermore, based on the length of the project, the subsidiary plans may be developed using progressive or rolling wave planning and may need to be updated on a regular basis. In fact, in view of the typically long durations of complex projects and uncertainties surrounding project end products, rolling wave planning, including both work packages and planning packages, will be a necessary and standard part of the schedule planning process.

PROJECT APPROVALS

Managing Traditional Projects	Managing Nontraditional Projects
The timing for project approvals is generally fast and does not create major issues with the schedule.	A sense of urgency among the stakeholders may not exist. Approvals may take a long time, and by the time the approval is made, you may be behind schedule.

The timing of approvals is critical for most project managers. On traditional projects with very few key stakeholders, the timing for approvals is generally very quick. Everyone seems to understand the need for urgency.

On complex projects, especially those involving developing countries, the sense of urgency as we know it may not exist. In some countries, decisions and approvals on complex projects may require the signature of several high-ranking ministers or government officials. When politics gets in the way of approvals, there is a slowdown in the approval process. Some key stakeholders may view the timing of the approvals as critical to their career and may invite the media to attend the actual signing.

When approvals can be delayed, the project manager must either allow the project to slip or assume the risk that the approvals will take place and continue on with the project. The risk with the latter approach is that, if the approvals do not take place, the project manager's company may be liable for the costs incurred. This is particularly important if long lead procurement is needed. To counteract some of these risks, contract provisions may specify time periods for approvals, which, if not met, release the contractor from liabilities for schedule slippage. Additionally, if possible, it's always best for the project manager to brief the key stakeholders, let them know where in the schedule they are responsible for making decisions, and advise them of the schedule consequences should those decisions not be made in a timely manner.

PROJECT'S CONSTRAINTS

Managing Traditional Projects	Managing Nontraditional Projects
The constraints on a project are reasonably well known, and their interaction may be predictable and controllable.	Handling complex constraints where the interactions may be unknown is very difficult.

With traditional projects, the constraints are reasonably well known and the interactions between the constraints are predictable. Trade-offs between competing constraints can take place on a regular basis, and relatively few stakeholders are needed to make a decision. Also, the prioritization of the constraints is known and agreed to by all of the players.

On complex projects, there can be multiple constraints, and the project manager may not even know about several of the constraints because they may be held in secret by key stakeholders. As we stated previously, not all of the stakeholders will agree to the project's goals and objectives.

The prioritization of the constraints is likewise difficult because getting stakeholder agreements may be difficult. The interaction among competing constraints may not be known because stakeholders may withhold information. If key stakeholders change during the project, the prioritization of the constraints can likewise change. The prioritization can also change according to the life-cycle phase.

In view of this, it's important for the project management team to regularly validate the prioritization among the competing constraints and to update them as changing priorities emerge. Since this prioritization informs the decisions made by project team members, they need to be informed through the communications process whenever significant changes in priorities among the competing constraints are made.

IDENTIFICATION OF DELIVERABLES

Managing Traditional Projects	Managing Nontraditional Projects
The deliverables are known and agreed to in the project management plan and the project plan.	Each stakeholder understands the deliverable in their group but may not fully understand the requirements from other groups. There may be a need to create a subsidiary plan just for the identification of deliverables.

Sometimes it is difficult to believe that there may be confusion over the project's deliverables. While this does not happen frequently on traditional projects, it does happen on complex projects. Stakeholders understand the deliverables within their group but may have a limited understanding of the deliverables in other stakeholder groups. The interaction between deliverables and the integration of those deliverables can also lead to confusion. For example, the deliverable from one stakeholder group could be an input requirement for a second stakeholder group. One stakeholder may believe that accepting a lower-quality raw material to be used in his/her deliverable will save money. The second stakeholder that must use the deliverable may find the quality of the preceding deliverable as unacceptable.

Previously, we stated that goals, objectives, and constraints can change due to politics, new stakeholders, and changes in the enterprise environmental factors. The same holds true for the deliverables. Getting stakeholders to agree on all of the deliverables may be an impossible task. Periodic review of the deliverables may be necessary. There may even be a need for a subsidiary plan just for identification of the deliverables, and the plan may need to be reviewed on a periodic basis.

In complex projects with iterative life cycles, the requirements that define the deliverables will frequently change. Each successive prototype will progressively refine the requirement set for these deliverables until an acceptable deliverable satisfies the customer's needs.

CHANGE MANAGEMENT

Managing Traditional Projects	Managing Nontraditional Projects
Change management systems are necessary for all projects. However, for traditional projects, decision making is reasonably fast.	It may be necessary to have a full-time person responsible for change management. The change management process is often slow because of the number of stakeholders and the fact that they often know very little about the technical details of the project and the real implication of their decision.

Change management systems are a necessity, more so to prevent unwanted changes than to approve necessary changes. On traditional projects with few stakeholders, the change management system is reasonably fast, since we can get stakeholder agreement. But on complex projects, getting stakeholder agreements may be difficult, thus slowing down the decision-making process.

Many of the stakeholders, even key stakeholders, may not be technically competent enough to make the technical decisions needed for approval or rejection of some scope change requests. While it is true that this can happen even on traditional projects, the large number of stakeholders on complex projects may need to get feedback from their own organizations where the sense of urgency does not exist.

The situation gets further complicated when even key stakeholders' organizations do not understand the technology behind the scope change request and therefore cannot determine the impact or interactions of an approved scope change on their organization, and yet they must cast a vote. Simply stated, the greater the number of stakeholders and the more complex the job, the more difficult it is to make scope change decisions in a timely manner.

CHANGE CONTROL MEETINGS

Managing Traditional Projects	Managing Nontraditional Projects
There may be very few meetings required of the change control board, and the meetings may be short since we are dealing with few stakeholders.	There may exist a need for continuous change control board meetings, and each meeting may be long in duration. It may also be difficult to get all of the participants to come to an agreement at the meetings.

Everyone seems to concur on the necessity for a change control process. Even with the best-planned projects, changes can occur. There are two ways to handle scope changes; one way is to make all of the necessary changes as they are discovered, and the second way is to delay making the changes until the project is over and then implement an enhancement project to make all of the changes after the primary contract has been completed.

On traditional projects, the timing is such that post-project enhancements may be an acceptable alternative. But on complex projects, waiting years or even a decade to make enhancements may be impossible and unrealistic. Therefore, complex projects generally suffer from the need for a continuous scope change control process. They may require regularly scheduled change control board meetings.

The high frequency of these meetings, coupled with the large number of stakeholders, can bring the project quickly to its knees if the stakeholders cannot fashion processes that allow them to reach change agreements quickly. Some key stakeholders may have other commitments and not be able to attend monthly change control board meetings.

One solution might be the establishment of a maximum size of the change control board, and attendance will be limited to those that are directly involved with or affected by the scope change. This approach, however, may alienate certain key stakeholders that wish to be involved in all of these meetings, especially if the approval of a scope change requires additional funding requests.

CONDUCTING MEETINGS

Managing Traditional Projects	Managing Nontraditional Projects
There is a well-understood schedule for team meetings, and getting the right people to attend is usually not an issue.	The management of a team meeting can be very complex. Getting the right people at the right time is critical. This can become more difficult if we are dealing with a virtual team.

The management of meetings is a crucial skill for all project managers. Project team meetings should be for the exchange of information and, when necessary, decision making. All too often, team meetings are viewed as the place where action items are initiated. While action items may be necessary in some cases, action items actually create unnecessary additional meetings.

There are numerous effective ways to conduct meetings. First, people should be provided with an agenda at least a few days prior to the meeting. If decisions will be necessary, then a description of each problem and the type of decision needed should be identified in the agenda so that the stakeholders can think about the problem prior to attending the meeting or invite some of their own people with the necessary authority or technology to attend.

Another way of making meetings more effective is to provide the attendees with copies of all of the meeting materials at least a few days prior to the meeting. Action items are the result of people feeling uncomfortable about making a decision right now and wanting to think about it a little longer. Expecting people to show up at a meeting, see the problem for the first time, and be expected to make a decision rapidly is wishful thinking.

PARTNERSHIPS AND ALLIANCES

Managing Traditional Projects	Managing Nontraditional Projects
Partnerships and alliances will remain intact throughout the duration of the project.	Partnerships and alliances can change throughout the project. This can cause major delays in meeting milestones.

On traditional projects, partnerships and alliances generally remain intact for the duration of the project. On complex projects, this may not be the case. Factors such as political conflicts (intergovernmental as well as intra- and interorganizational), changes in government administration, culture, and religion can force the dissolution of alliances and the creation of new alliances.

It may be unrealistic to assume that these alliances will remain for the duration of the project. New partnership agreements may need to be forged. The final result can very well be an elongation of the schedule and cost overruns.

The project management team has little control over the forging and dissolution of these alliances and partnerships. Project management issues that derive from these changes can affect all nine *PMBOK®* *Guide* areas. While some of these changes are made in order to have a positive impact on specific project constraints (changing alliances to promote easier access to project funding, for example), consequential negative effects are inevitable on other constraints (schedule, risk, resources, etc.).

ABILITY TO CHANGE

Managing Traditional Projects	Managing Nontraditional Projects
Companies may not find it necessary to adapt to any type of change as a result of the completion of the project.	Companies must be willing to change, even though it may be a struggle.

On complex projects, there is a greater likelihood that the completion of the project will result in some change in the way the company does business. Not all stakeholders may be happy about changing the way they do business. For some stakeholders, the change may result in a reduction of their empire, loss of authority or power, reduction in status, or even a loss of employment. As discussed previously, not all stakeholders may wish to see the project completed successfully, even though they appear as key stakeholders and demonstrate verbal support for the project.

What this all means is that project managers on complex projects must also have the skills of a transformational leader. The project manager must do a thorough stakeholder analysis (see Chapter 10), identifying those stakeholders in whom resistance to change is most likely to occur, and develop strategies to neutralize opposition to, and reinforce support for, the changes that will result from the project. This is not an easy task, particularly if the effects of the project represent a significant departure from current operations, if they are widespread, or both.

We humans have evolved to be wary of change. It can affect us in many ways, resulting in feelings of denial, anger, sadness, disorientation, and depression.[2] If change in an organization is managed properly, these feelings can be minimized, but never completely eliminated.

---

[2] See William Bridges, *Managing Transitions*, 3rd ed. Cambridge, MA: Da Capo Press, 2009.

3

SCOPE
MANAGEMENT

PROJECT BOUNDARIES

Managing Traditional Projects	Managing Nontraditional Projects
The boundaries of the project are reasonably well known and realistic.	The boundaries of the project can and will change based on a multitude of factors such as a changing environment or changes in management.

We all take for granted that the boundaries of the project are known and well understood. While this may be true on traditional projects, it is certainly not true on complex projects. The longer the project, the greater the likelihood that the boundaries will change.

There are several reasons for this. First, the stakeholders involved in planning the complex project may have very limited expertise with this type of project as well as with the establishment of boundaries. Second, as discussed in Chapter 2, stakeholders are more likely to establish optimistic rather than pessimistic boundaries, even when they understand the limitations to their knowledge. Third, each stakeholder will set boundaries according to their own personal interests and desires, often with little regard for the concerns of other stakeholders.

There are, of course, other factors that can cause boundaries to change other than stakeholders' desires. These factors include changes in government administration, politics, stakeholder agreements that cannot be enforced, and changes in the enterprise environmental factors.

STAKEHOLDER IDENTIFICATION

Managing Traditional Projects	Managing Nontraditional Projects
The project's key stakeholders are easily identified, and their numbers are limited and manageable. Developing a stakeholder register is a relatively straightforward effort.	On many complex projects, particularly on those of extended duration and on projects in which multiple numbers and types (industry, government) of organizations are involved, key stakeholders are numerous and not easily identified. The development of the stakeholder register may involve much greater effort than in a traditional project.

Stakeholder management begins with stakeholder identification. This is easier said than done, especially if the project is multinational. On traditional projects, the number of stakeholders is relatively small and easily managed by the project manager. Developing a stakeholder register is therefore relatively straightforward. This is not so on complex projects. Stakeholders can exist at any level of management. Corporate stakeholders are often easier to identify than political or governmental stakeholders.

Each stakeholder is an essential piece of the piece of the project puzzle. Stakeholders must work together and usually interact with the project through the governance process. Therefore, it is essential to know which stakeholders will participate in governance and which will not. This is one of the reasons why a stakeholder register may be essential on complex projects.

As part of stakeholder identification, the project manager must know whether he or she, as the project manager, has the authority or perceived status to interface with the stakeholders. Some stakeholders perceive themselves as of a higher stature than the project manager, and, in this case, the project sponsor may be the person to maintain those interactions.

REQUIREMENTS COLLECTION

Managing Traditional Projects	Managing Nontraditional Projects
The output (product, service, result) of the project is well defined based on an established set of end-user requirements.	Particularly in research and development (R&D) and new product development projects, the output has never before been seen. Therefore, it is difficult to associate specific requirements with the product until prototypes have been developed and end-user feedback has been obtained.

On traditional projects, the outputs or the deliverables of the project may be well defined. On complex projects, with a large number of stakeholders, many of whom may have limited knowledge of the project or the technology, requirements collection can be daunting. In such cases, you may find that the stakeholders will define project success according to time and cost rather than the accomplishment of the requirements and the quality of the results.

Requirements collection, along with project planning, may have to be accomplished using progressive planning or rolling wave planning. As an example, the final requirement set may not be able to be defined until several successive prototypes have been built and stakeholder feedback and approval has been obtained. It is important to remember that, while traditional projects with the same general types of outputs may have some degree of similarity and reasonably good estimating techniques, complex projects suffer from significant uncertainty, perhaps very little similarity to any other projects, and having a moving target for an output.

On some complex information technology (IT) projects, it is possible for the requirements to change because of changes in business conditions such as consumer preferences, changes in technology, and the implementation of newer processes.

CHANGING PRODUCT REQUIREMENTS

Managing Traditional Projects	Managing Nontraditional Projects
The product requirements on which the project scope is based are stable and well defined.	Due to changing market conditions (e.g., competition, end-user perceptions of need, technology changes), particularly in projects of extended duration, product requirements are in flux throughout much of the project. Consequently, project scope needs to commensurately change.

The product requirements on traditional projects are usually well defined and stable for the duration of the project. The assumption is usually made that technology is known and will not change over the duration of the project. Projects less than 12 to 18 months in duration are often considered to fall into this category.

On nontraditional projects, the duration of the project can play havoc with predetermined product requirements. Factors mentioned previously, such as changing customer needs, implementation of new processes, and new technologies may force changes in product requirements such that obsolete deliverables are not produced.

There comes a point on all projects, whether traditional or nontraditional, when one must decide to eliminate "creeping elegance" and launch the product. All additional change requests to the product's requirements may need to be completed with an enhancement project to create the next generation of the product. If the complex project is to create deliverables to be sold in the marketplace, then continuously changing product requirements may force the selling price of the product to be overpriced in the marketplace.

THE PROJECT PLAN: WORK PACKAGE LEVELS

Managing Traditional Projects	Managing Nontraditional Projects
The project plan may be well defined through the detailed levels of the work breakdown structure (WBS) including the work packages.	While feeble attempts are made at the initiation of the project to produce the work packages, the work packages are normally created using some form of rolling wave planning or progressive planning.

On traditional projects, the project plan is usually well defined through the detailed levels of the work breakdown structure (WBS), including work packages. Sometimes, project managers with a command of technology tend to force the work packages to the extremely detailed levels of the WBS when, in fact, the project could be managed at much higher WBS levels.

On complex projects, the reverse is true. Project managers tend to want to manage at too high a level of the WBS. There are several reasons for this. First, at project initiation, the only levels of the WBS that may be known with reasonable certainty are the management levels (i.e., the top three levels of the WBS). Therefore, we end up with high-level work packages. Second, not all of the team members will have sufficient skills such that more detailed work packages can be created, and the project manager, as well, may have limited knowledge. Third, even if work packages can be developed, they most likely will be the work packages for the first three to six months of the project. Progressive or rolling wave planning may be necessary for further development of the next set of work packages.

PROJECT'S DELIVERABLES

Managing Traditional Projects	Managing Nontraditional Projects
The deliverables of the work packages are reasonably well understood and predictable.	The deliverables of the work packages may end up surprising the project team and users.

Previously, in Chapter 2, we stated that it is sometimes difficult to believe that there may be confusion over the project's deliverables. While this does not happen frequently on traditional projects, it does happen on complex projects. Stakeholders understand the deliverables within their group but may have a limited understanding of the deliverables in other stakeholder groups. The interaction between deliverables can also lead to confusion. For example, the deliverable from one stakeholder group could be an input requirement for a second stakeholder group. One stakeholder may believe that accepting a lower-quality raw material to be used in his or her deliverable will save money. The second stakeholder that must use the deliverable may find the quality as unacceptable.

While the deliverables of the work packages are reasonably well understood and predictable on traditional projects, the same cannot be said for complex projects. The deliverables of the work packages in complex projects may end up surprising the project team, the users, and the stakeholders. Usually, the surprise is greeted unfavorably rather than favorably. Unfavorable surprises can result cancellation of the project, redirection of the project by key stakeholders, a change in leadership of the project, and possibly greater governance by the stakeholders for the remainder of the project.

WORK PERFORMANCE INFORMATION

Managing Traditional Projects	Managing Nontraditional Projects
Work performance information is gathered using a single, automated time reporting system. The data from that system provide the project manager a window on how much of the scope has been accomplished and how much remains.	Work performance information is much more difficult to obtain due to the numbers of subcontractor and partner organizations involved. In some cases, no time reporting systems may be available; and multiple time reporting systems may use incompatible data collection formats. In other cases, time reporting may not be done at all.

In Chapter 2, we discussed that organizational project assets include the tools available to the project manager for managing the project. The tools may be corporate policies, procedures, guidelines, forms, templates, and checklists as well as the EPM systems itself. On traditional projects, all of the tools usually reside in the contractor's organization and the tools, most of which are automated, provide information on work performance. Three typical reports are (1) progress reports, (2) status reports and (3) forecast reports.

On complex projects, each stakeholder may be at a different level of project management maturity and therefore possess a variety of different project management tools, many of which may be obsolete or simply not capable of interfacing with the project manager's tool box. Some of the tools may not be automated and may not be compatible with the work performance measurement system used by the project manager.

Project managers may wish to track costs at the work package levels and establish charge numbers for each work package. However, part of the project team may be using a legacy system that is not capable of defining the detail of the WBS down to the work package level. It may, therefore, not be possible to ultimately understand the accuracy of detailed estimates by comparing original work package estimates with their actual values due to a lack of this detail.

VERIFY SCOPE

Managing Traditional Projects	Managing Nontraditional Projects
Each element of the WBS has been defined. The discrete work packages all have specific artifacts (deliverables) and associated completion criteria.	The WBS may have a preponderance of planning packages, rather than work packages, due to the ambiguous nature of the work. In cases where prototype development is required, the artifacts and completion criteria are self-defining based on end-user acceptance.

During project execution, once the work on a deliverable (or artifact) has been finished, tools/techniques of the Verify Scope process are used to ensure the deliverable's acceptability to the sponsor or client. This applies to interim deliverables as well as to the final product. (This should not be confused with the quality control process that ensures the accord of the deliverable's attributes with predefined quality specifications.) In traditional projects, where requirements and deliverable attributes are well defined and understood at the beginning of the project, acceptance should be a fairly routine, if not formal, procedure. The tools used to accomplish this verification can range from a test mask (used in validating performance of electronics deliverables) to a formal document review.

As noted previously, product requirements and associated deliverable attributes may not be defined until late into project execution. This is particularly true of projects using (or developing) new technologies and research and development projects. In these types of projects, the attributes of both interim deliverables and end products may be refined throughout the course of project execution and may not finally be defined until an "acceptable" end product is produced (see Chapter 8 on "satisficing" zones). In these complex projects, iterative prototype development will necessarily require the client to review and give feedback on each successive prototype, thereby involving the client in the iterative definition of the end product's requirements.

We may also see complex projects in which multiple end-product options are explored simultaneously, with the option that produces an end product closest to meeting the current business needs selected as the final end product. An example of this (although not necessarily managed as a "project" at the time, and not producing an end product with wide market acceptance) occurred in Digital Equipment Corporation (DEC) in the late 1970s. In response to the advent of the early IBM and Apple personal computers, multiple engineering teams were given the task of developing a PC that would meet DEC's client base's demands. The end result was the DEC Rainbow personal computer.

CONTROL SCOPE

Managing Traditional Projects	Managing Nontraditional Projects
The scope of the project requires minimal change. When changes are required, they are handled by the project manager and project team using a well-defined change control process.	The scope of the project changes frequently necessitating the need for a more flexible change control process. However, the involvement of significant numbers of stakeholders tends to make the decision making around scope changes more complex and time consuming, generally resulting in a more formal (change control boards) and less flexible process.

Everyone seems to concur on the necessity for a change control process. This was discussed in Chapter 2. Even with the best-planned projects, changes can occur. With possibly a high frequency of meetings and a large number of stakeholders, the scope change control process can bring a project to its knees if the stakeholders cannot come to an agreement quickly. Some key stakeholders may have other commitments and not be able to attend monthly change control board meetings, while other stakeholders would prefer to make all of the decisions by themselves. This is a more informal approach than a change control board meeting and takes less time, but does run the risk that critical information from other stakeholders will not be available.

One solution might be the establishment of a maximum size of the change control board in which attendance would be limited to those who are directly involved with or affected by the scope change. This approach may alienate certain key stakeholders who wish to be involved in all of these meetings, especially if the approval of a scope change requires additional funding requests.

4

TIME

MANAGEMENT

PROJECT DEPENDENCIES

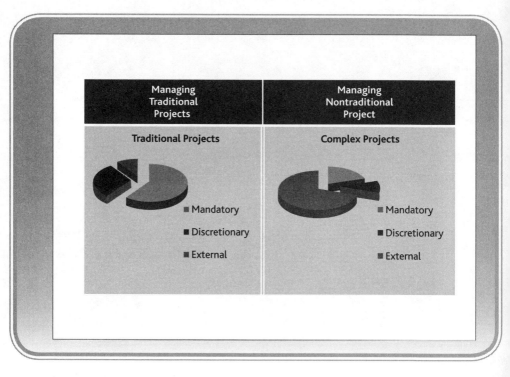

On traditional projects, where most of the work is done in house with existing resources that may be under the control of the project or line managers, the majority of the dependencies are mandatory dependencies where the relationships between work packages are well known. Some dependencies may be discretionary dependencies such as the relationship between procurement activities and producing the bill of materials. The only external dependencies, if they exist at all, would be the contractors that perform work that affects activities in the precedence diagram, or activity dependencies from other projects in the program of which the project is a part.

On complex projects, perhaps as many as 90 percent or more of the dependencies may be external dependencies such as stakeholder participation, sign-offs by stakeholders, politics, culture, and other such issues. The interactions between the work package may be complex such that the project manager may simply assume that all of the dependencies are external dependencies.

TEMPLATES

Managing Traditional Projects	Managing Nontraditional Projects
The smaller and less complex the project, the more likely that time management templates can be used effectively.	It is highly unlikely that all of the players will be using the same templates for time management. Each player may have a different scheduling technique.

Traditional projects are usually managed by an enterprise project management methodology, which may be based on predefined templates, forms, guidelines, and checklists. The purpose of the enterprise project management (EPM) system is to provide the project manager with some degree of standardization and control for the management of the project.

Complex projects where each stakeholder may or may not have any project management methodology, and where the methodologies are all different if they exist at all, makes the work of the project manager very difficult. Expecting each of the stakeholders to readily accept the use of the project manager's preferred tool set may be unrealistic.

Each stakeholder may desire the use of a different tool for scheduling work, and some stakeholders may even want it done manually. Getting all of the team, regardless of where they are located, to agree to a uniform set of tools would be ideal but impractical. Even if they did agree to a common set of tools, the time needed to train them all on the use of the tools might be prohibitive.

ACTIVITY LIST

Managing Traditional Projects	Managing Nontraditional Projects
On traditional projects, the activity list is relatively small, and project managers may even have work packages at the lower levels of the work breakdown structure (WBS).	There will be a compromise between high-level and low-level WBS schedules because of the number of activities in the WBS. Not all WBS activities will appear in the schedules.

Some project managers are fortunate enough to have EPM methodologies that have templates for the activities that reside in usually the top three levels of the WBS. There may also be some templates for work package activities if there is a great deal of commonality between projects. All of this is possible for traditional projects, but not so for complex projects.

If the project manager does not possess a command of technology on a traditional project, then the project manager works closely with the functional managers to prepare a detailed activity list down through the work package levels. The functional managers then assume accountability for management of the work packages. And even then, the activity list is relatively small.

On complex projects, there may need to be a compromise between high-level and low-level schedules. First, it may be impossible for the project manager to work with all of the players and prepare a detailed list of activities down to the work package level. And even if this could be done, the number of work packages and subsequent complexity of the detailed schedule would negate the project manager's ability to effectively manage the project.

A compromise between high-level and low-level schedules is needed. The compromise may be determined by the amount of project management knowledge and experience each team member possesses as well as their company's tool kit for their project managers.

PROJECT SCHEDULE

Managing Traditional Projects	Managing Nontraditional Projects
People have faith in the project schedule and believe it to be a realistic schedule.	Because the estimating function may be poor, the schedule is generally unrealistic and subject to change.

On traditional projects, the number of people involved in schedule preparation is relatively small and it is usually easy to get them to have ownership in the plan even if the plan is somewhat optimistic. Most of the time, realistic plans are developed and few changes are made that affect the end date.

Complex projects notoriously suffer from schedule slippages and cost overruns due to poor estimating. It is unrealistic to assume that all of the players have good estimating techniques. With traditional projects, generally there may exist only a small group of project estimators to support the project. On complex projects, each stakeholder group can have their own group of subject matter experts, but not all subject matter experts are the same even within the same technical discipline. It is possible that a senior estimator in a developing nation would have only the skills of a junior estimator in an industrialized nation.

With a multitude of subject matter experts, it may be very difficult to get agreement on the final schedule. Some subject matter expert groups may be heavily biased by their stakeholders or superiors to promote a schedule that is not in the best interest of the entire project. Politics, culture, religion, and perceived status can influence the decisions of these groups.

PURPOSE OF SCHEDULE

Managing Traditional Projects	Managing Nontraditional Projects
The schedule is used as a means of project control.	The schedule is used as a communication tool, to get buy-in from stakeholders and to motivate team members.

Schedules can mean different things to different people. On traditional projects, schedules are used as a means of controlling the project. On complex projects, the schedule serves more as a communications tool to get buy-in from the stakeholders, keep them informed and also as a means of motivating team members.

On traditional projects, we tend to track the details of the schedule and continuously look for ways to compress the schedule. On complex projects, stakeholders focus on the deliverables and accomplishment of key milestones rather than lower-level work package.

The purpose of the schedule can also be affected by the tools available at each stakeholder's location. Stakeholders with good scheduling tools may track the schedule more closely than other stakeholders.

Some project managers have found success in using time-phased precedence diagramming technology with their project teams and, at times, other stakeholders. These tools may be particularly helpful in gaining buy-in to the schedule and emphasizing the importance of completing various types of activities on time (e.g., critical path activities, high-risk activities, reviews, and approvals).

TYPES OF SCHEDULES

Managing Traditional Projects	Managing Nontraditional Projects
One type of schedule will suffice for the entire project. However, status reporting to the customer may take place with bar charts and milestone charts.	Multiple scheduling techniques will be used. Based on the maturity level of the partners, some may still be using rather unsophisticated charting techniques.

There are three types of schedules; bar (Gantt) charts, milestone charts, and networks. Network diagrams are the most commonly used scheduling techniques because they serve as early warning systems identifying downstream problems associated with upstream delays in work packages. Traditional projects may have the option of up to 250 scheduling software packages to select from. However, companies that develop EPM systems use one, and only one, scheduling system across the entire company.

On complex projects, multiple scheduling techniques are a necessity. This, of course, is based on the project management maturity level of the partners. Some partners may be quite comfortable with rather unsophisticated scheduling techniques, while others may require the use of network techniques.

And while project managers will manage the project with network-based techniques, stakeholder reporting is usually accomplished with bar charts and milestones charts, where the information can be more easily displayed and explained.

PUBLISHED ESTIMATING DATA

Managing Traditional Projects	Managing Nontraditional Projects
On certain types of traditional projects, there is an abundance of published estimating data.	For nontraditional projects, the amount of published estimating data is usually at a minimum. Partners, particularly in developing markets, may have limited access to any published data for estimating.

Project teams in certain industries have the luxury of being able to rely on published databases for estimating. Construction estimating, where a long history and well-compiled databases exist, would be an example. On complex projects, the amount of published (or even compiled) historical data may be minimal. One of the reasons for that is that the estimating data on complex projects is viewed as a competitive weapon, and companies are unwilling to share their information.

Partners on complex projects may have limited access to estimating data, especially if they are from developing market countries. Here in the United States, there are numerous seminars on construction cost estimating, software development estimating, and other such courses. In developing markets, workers may not have access to these seminars or this information.

What does this mean for the project? There are three levels of accuracy associated with activity estimates: single-point estimates (the least accurate), range estimates (somewhat more accurate than single-point), and estimates based on collected actual data from similar activities in previous projects (the most accurate). If historical estimates are unavailable, the project team will have to develop single-point or, better still, range estimates for the project's activities.

PROJECT MANAGEMENT SOFTWARE

Managing Traditional Projects	Managing Nontraditional Projects
Traditional projects are usually managed with one, and only one, software package for scheduling.	On nontraditional or complex projects, there may be several different types of scheduling tools, and some of the tools may be incompatible with those used by their partners.

On traditional projects, we normally have the opportunity to select from a multitude of software packages, although most companies require that their projects be managed with only one software package for standardization and consistency purposes. On nontraditional projects with virtual teams, many of which come from developing market nations, there can be a large number and type of scheduling tools and techniques that are in use.

Many of these techniques may be outdated but are still being used by some of the partners. Many of the packages may be incompatible with the techniques used by their partners. The cost of selecting one standard scheduling technique for all of the players, combined with the cost of training, will most likely not be cost effective.

We should note, however, that there may be tools (or companies) that will translate data from one application for use in another (e.g., translating data from Microsoft Project for use in Primavera). In the absence of this translation, the various types of schedule data will have to be gathered by the project schedulers, analyzed type by type, and put into an overall project schedule that can then be used for status updates and progress reports to stakeholders and project team members. This approach involves much more work for the project schedulers, and the risk of mistranslation of schedule data, since it's being done manually, is much higher. Clearly, this is not the preferred method of updating schedules. However, if the project partners are not willing to invest the time and effort into selecting a single scheduling tool and training everyone in its use, it may be the only method available.

There are more frequent instances now in which the project owner's request for proposal (RFP) specifies the necessary use by all contractors of a single tool for schedule reporting. How widespread that becomes remains to be seen.

TOP-DOWN VERSUS BOTTOM-UP ESTIMATING

Managing Traditional Projects	Managing Nontraditional Projects
Estimating can be top-down or bottom-up, although top-down is usually the preferred approach.	On nontraditional or complex projects, bottom-up estimating is rarely used since it requires a good knowledge of the work to be accomplished. Some strategic partners may have a limited knowledge of their own role on the project and their responsibilities.

On traditional projects, we can estimate the work either top-down or bottom-up. Top-down is the preferred approach in general, but projects in manufacturing or projects where really good detailed estimates exist may use bottom-up estimating. Bottom-up estimating requires an exceptionally good knowledge of the work to be done.

On complex projects, there is no guarantee that the resources in the partners' companies possess this knowledge. Strategic partners and key stakeholders may not even understand their own role on the project.

THREE-POINT ESTIMATES

Managing Traditional Projects	Managing Nontraditional Projects
Three-point estimates are commonly used, provided that we can realistically estimate the optimistic and pessimistic extremity points.	Three-point estimates may not be appropriate or applicable. And even if they are appropriate, the risks that will be identified using the three-point estimates may be so large that the project will be canceled or removed from the queue.

The purpose of three-point estimates is to establish a range for the estimates such that the risk associated with individual activities in the schedule can be assessed. If the optimistic and pessimistic estimates for the duration of work are far apart, then the assignment of the risks will be meaningless.

If strategic partners use optimistic and pessimistic estimates without really understanding what they are doing, there is a high probability that the project may be canceled. The corollary is also true. Some partners may not want to see the project canceled and may therefore establish tight boundaries around their optimistic, most likely, and pessimistic estimates such that the project's schedule risk is minimized for estimating purposes. Then, some time into execution of the project, we discover that the estimates are, in fact, wrong, and again key stakeholders can have visions of large cost overruns and schedule slippages.

In any case, the project manager needs to thoroughly understand the parameters around project estimates. If these estimates are thought to be unrealistic, the project manager needs to surface these concerns with the stakeholder involved. If the stakeholder maintains the validity of the estimates, despite clear indications otherwise, the project manager needs to account for this in the risk management process.

DURATION VERSUS EFFORT

Managing Traditional Projects	Managing Nontraditional Projects
On traditional projects, we tend to have a clear understanding of the difference between duration and effort, and we usually assume that all effort begins as early as possible within the duration.	On complex projects, neither effort nor duration may be known with any degree of certainty. With some partners, the ability to define effort may be quite poor because the skill level of the workers is not known with any degree of certainty.

On traditional projects, we seem to have a clear understanding of the difference between duration and effort. To minimize project risks, especially schedule risks, we tend to assign the resources as early as possible so that slack remains in the schedule in case rework is needed.

On complex projects, neither duration nor effort may be known with any degree of certainty. This often occurs because the strategic partners are unsure about the skill levels of the workers to be assigned to the project. Some assumptions must be made about these skill levels so that some sort of schedule can be developed. These assumptions can be vetted in the risk management process, and appropriate contingencies can be addressed at that time.

Another potential problem occurs when the strategic partners practice backward planning (rather than forward planning), thereby ensuring that activities will start at the latest possible time. This practice eliminates slack from the project's critical path, thereby ensuring a late delivery of interim deliverables and the end product.

"WHAT-IF" SCENARIOS

Managing Traditional Projects	Managing Nontraditional Projects
What-if scenarios are common practice for the development of the risk management plan and the establishment of reserves and contingencies.	What-if scenarios are avoided because the identification of the risks may result in the project's not being approved or canceled even after it begins. What-if scenarios can bring forth information that people prefer to have hidden.

The purpose of "what-if" scenarios is to anticipate what can go wrong, perform risk management in a timely manner, and develop the necessary contingency plans and reserves. This brings risks to the surface so that they can be addressed in a timely manner.

On complex projects, bringing risks to the surface or even discussing them openly, can lead to disaster, project cancellation, or replacement of the workers who brought forth the risks. Quite often, government officials in emerging-market countries approve projects without considering the risks. Risk exposure is often hidden or simply not discussed until after project go-ahead.

SCHEDULE COMPRESSION TECHNIQUES

Managing Traditional Projects	Managing Nontraditional Projects
Project managers are familiar with the various techniques for schedule compression. The techniques are used whenever possible to accelerate the completion of the project.	Schedule compression techniques may be avoided if the workers view the project as job security, or if the corporate culture discourages employees from working faster.

On traditional projects, the project manager is usually willing to use any or all of the five basic schedule compression techniques, namely, overtime, crashing, scope reduction, outsourcing, or parallelization. Each of these techniques bring with it advantages and disadvantages. However, on complex projects, the disadvantages may outweigh the advantages. For example:

- *Working overtime*. Some countries discourage working overtime because the overtime pay could result in a new "class" of workers.

- *Crashing*. Crashing implies that additional internal resources are available to perform the work. This might not be the case, and even if it could be done, the skill level of the additional resources may be unacceptable.

- *Scope reduction*. Scope reduction can work if, and only if, the scope can be eliminated without destroying the integrity of the project. Complex projects generally do not have excess scope that can be eliminated.

- *Outsourcing*. Outsourcing can work only if qualified suppliers can be found. The risk is that you are creating additional external dependencies.

- *Parallelization*. Performing work in parallel rather than in series may seem a worthwhile risk. However, if rework is required, as is always the case on complex projects, multiple work packages can be affected, thus elongating the schedule.

Chapter

5

COST
MANAGEMENT

THE BASIS FOR PROJECT FUNDING

Managing Traditional Projects	Managing Nontraditional Projects
Project funding is based on an agreed-upon detailed plan. If the plan changes, funding can change to fit the plan. This can occur throughout the project life cycle.	Project funding is based on a high-level milestone plan that is agreed to by the stakeholders. Changes in the high-level plan will mandate a rejustification of both the plan and the accompanying budget. Additional funding, if needed, may not be available until the next funding cycle.

On traditional projects, the funding for the entire project is esta-blished at the beginning of the project and usually remains the same over the project's life cycle unless the plan undergoes scope changes. For nontraditional projects, the funding may be based on the completion of high-level milestones.

The larger the project, the greater the tendency is for the project to be funded according to a funding cycle, such as a yearly funding cycle. Scope changes have to be justified and the overall project rejustified, but funding for the scope changes may not be available until the next funding cycle. The consequences of this are signifi-cantly detrimental to the project schedule and can ultimately increase the overall cost of the project.

Most project team members are assigned to work on more than one project at a time. Companies can't just let their resources sit idle while waiting for funding decisions to be made. So those working on a project that has temporarily stopped will probably be reassigned to other ongoing projects. Even after the project's incremental funding has been approved, it may prove difficult getting the resources back onto the project. And those who are able to return may experience productivity deficits trying to pick up where they previously left off. All of these effects lead to increased costs on the project.

PROJECT FUNDING

Managing Traditional Projects	Managing Nontraditional Projects
Project funding is relatively stable for the duration of the project unless scope changes are approved.	Project funding is on a yearly basis and can be unstable based on politics, the economy, withdrawal of support by certain stakeholders, and the inclusion of higher-priority projects into the portfolio of projects.

Traditional projects generally have stable funding. Once the project is approved and funding is established, the funding exists for the entire life cycle of the project, which is usually 18 months or less. No funding cycles are necessary. All scope changes will go through a scope change control process, and reserves are usually set up knowing that scope change requests will most likely happen.

On most complex projects, funding cycles are used. The most common funding cycle is a yearly funding cycle, but the partitioning of the money may be made according to quarterly review meetings, completed milestones, or other such activities.

Complex projects are known for unstable funding cycles due to politics, economic conditions, or changes in high-level administrative positions. Even though the complex project can be progressing according to plan and all stakeholders are pleased with the results thus far, higher-priority projects can suddenly appear in the portfolio of projects, and this project may be downsized, temporarily delayed, or canceled.

MULTIPLE FUNDING SOURCES

Managing Traditional Projects	Managing Nontraditional Projects
Project managers generally have one, and only one, funding source for the project. This allows for rapid decision making involving scope changes.	Project managers must deal with multiple funding sources, each with a different priority. Decision making is slow, and conflicts arise as to which funding source(s) will pay for the scope changes.

On traditional projects, there is generally one, and only one, funding source. This allows for rapid decisions involving scope changes because of the relatively small size of the change control board.

On nontraditional projects, the project team may have to deal with multiple funding sources, each with a different view of the priority of the project. Multiple funding sources can make the scope change control process difficult. Not all stakeholders will agree to the necessity for the some scope changes, and even if the scope changes are approved, there can be disagreements as to who will pay for them.

Another critical issue involves when the funding for the scope changes comes from stakeholders who work off of funding cycles. Even though the stakeholders can come to an agreement on the urgency of the scope change, money may not be available until the beginning of the next funding cycle—and that assumes higher-priority projects do not come along. The greater the number of funding sources, the greater the likelihood that the project's schedule will slip.

MANAGEMENT RESERVES

Managing Traditional Projects	Managing Nontraditional Projects
Usually, there exists one, and only one, management reserve to be controlled by the project manager and used for escalations in salaries, overhead rates, and procurement.	There may exist several reserves, each controlled by the partners. The reserves may be retained in secret for fear that disclosure may indicate risks that could result in project cancellation.

Contingency reserves in project budgets are established to implement risk response strategies for identified, high-severity project risks. Management reserves were originally established to compensate for unforeseeable risk, escalation factors in salaries, overhead rates, and procurement costs. Contingency reserves are normally under the control of the project manager, while management reserves are controlled by the project team's management. The client also establishes a management reserve, knowing full well that its requests for scope changes are inevitable and that the project manager's contingency reserves will not pay for these scope changes.

On complex or nontraditional projects, each funding source may or may not establish its own reserve. Some stakeholders may view the need for a reserve as an indication that not all of the project's risks were disclosed. This could lead to a cancellation of the project or the failure to approve needed scope changes.

Management reserves can compensate for stakeholders that may be limited according funding cycles for scope change approvals. Unfortunately, not all partners will maintain a reserve, and the project manager cannot force the partners to do so.

COST-ESTIMATING TECHNIQUES

Managing Traditional Projects	Managing Nontraditional Projects
Each functional unit may have their own estimating techniques but, for the most part, they are based on historical standards and are reasonably reliable.	Multiple estimating techniques can exist, and many could be just seat-of-the-pants estimates. Most partners may not have an estimating group and may have very little in the way of historical estimating databases.

On traditional projects in companies experienced in project management, many different estimating techniques can exist. Some companies maintain an estimating group that has historical data from which to make estimates. Usually, traditional project estimating is reasonably accurate for most industries.

On complex projects, partners may be relatively immature in project management, have limited experience in the various estimating techniques, have no historical databases, and have no estimating group. Therefore, the project manager may not know the quality of any of the estimates that are provided by the partners or various stakeholders. In some of the partners' companies, the estimates may have even been provided by people not affiliated with the project.

USE OF EARNED VALUE MEASUREMENT

Managing Traditional Projects	Managing Nontraditional Projects
Earned value measurement is being used, but perhaps not all of the components.	Software packages are being used for schedule management only. Earned value measurement may not be used because the company has no way of capturing the required data.

Earned value measurement is perhaps the best system for controlling a project. Most project managers are familiar with earned value measurement, although some companies are reluctant to implement it. Earned value measurement provides guidance on identifying variances from a plan or baselines, determining the cause of the variance, developing a corrective action, and measuring the new variances against the baselines. Most project management software contains the earned value measurement formulas.

On complex projects, software is used primarily for schedule management and information reporting. For earned value measurement to be effective, cost data must be accurately captured. Partners and stakeholders in organizations that are relatively immature in project management may have no means of capturing the necessary data, most of which may be based on labor hours expended in project execution. The project manager may then find that some of the partners can provide the necessary earned value measurement data and some cannot provide it. The result may be that the project manager may need to integrate cost data from multiple sources, many of which may be incompatible with the project manager's systems. This could negatively affect the quality of the financial and schedule data that the project manager is required to provide to the stakeholders.

FORECAST REPORTS

Managing Traditional Projects	Managing Nontraditional Projects
Forecast reports are being used and include estimated cost at completion, estimated time at completion, and other such forecasts.	Forecast reporting is avoided because of the risks. People are afraid to expose the reality of the progress for fear that the project may be canceled.

orecast reporting generally provides data on what the time and cost will be at the completion of the project. Today, we are also forecasting what the benefits and value will be at completion. The formulas for time and cost at completion are part of the previously discussed earned value measurement system that is part of most project management software packages.

One of the intents of forecasting time and cost at completion is to expose the reality of the progress such that corrective action can be taken if necessary. On traditional projects, we generally take corrective action rather than cancel the project. But on complex projects, exposing the risks of a schedule slippage or cost overrun early on in the project may create the fear in the minds of the stakeholders that the situation can get progressively worse and that the project should be canceled. Not all of the stakeholders have a good understanding that cost overrun or schedule slippage projections detected early in the project can be corrected.

It should also be noted that historical data has shown that once a project is around 25 percent completed, if the project is running behind schedule and over the budget, the likelihood of meeting initial milestone targets is marginal. In fact, to forecast *best-case* budget outcomes, U.S. government agencies divide the Budget at Completion by the Cost Performance Index (CPI) × Schedule Performance Index (SPI). So organizations must assess the forecasts and make go/no-go decisions on current business conditions and realistic project value objectives.

Chapter

6

HUMAN
RESOURCES
MANAGEMENT

FERVENT BELIEF

Managing Traditional Projects	Managing Nontraditional Projects
Based on the length of the project, and the fact that employees may be working on multiple projects, it may be impossible to obtain a vigorous pursuit of the vision.	A fervent belief will permeate the entire project. Team members will be hired based on their fervent belief in the success of the project.

When people work on traditional projects, they may also end up spending part of their time on other projects as well. But on long-term, complex projects, people may be committed to just one project. If people are not motivated on the one project they are working on, then the project may be doomed to fail.

To alleviate the problem with motivation, project managers often try to get team members to have a fervent belief in the success of the project. While a fervent belief may appear good, it comes with the downside risks of not wanting to hear bad news, not wanting to report bad news, not wanting to admit failure, and believing that failure of the project could damage one's career and reputation.

There is no question that creating a fervent belief can help drive a project to success. But the downside risk is that fervent beliefs also prevent projects that should be canceled from being canceled. People may do anything possible to have the project continue, in hopes of a miracle, rather than admit to failure or run the risk of not being considered a "team player" with all the associated consequences thereof.

CONFLICTS OVER OBJECTIVES

Managing Traditional Projects	Managing Nontraditional Projects
There is agreement among the players on the objectives of the project, and the final agreement is documented in the project charter and project scope statement.	Conflict over the objectives can occur at any time, even though there may have been an initial agreement among the stakeholders at project initiation. The worst case is when the host and the project team disagree on the project's objective.

On complex projects, business case development and final establishment of the project's objectives may be a long, tedious process involving many stakeholders. Many stakeholders may succumb to accepting a poor business case with unacceptable objectives and later push for changes to be made. Not all stakeholders are equal when it comes to preparing a business case and establishing objectives on a complex project. There is also the possibility that certain key stakeholders who should have been involved in setting project objectives have, in fact, not been.

The length of the project can also create conflicts over objectives. In the initiation phases of a complex project, there is a tendency to provide lip service toward an agreement of the objectives just to make sure that the project is kicked off. Later on, the conflicts will come to the surface, and scope changes may be necessary.

Conflicts over objectives can occur if there are changes in the key players or governance committees. Not all players will establish objectives in the best interest of the project or the client. In many instances, the conflicts involve self-serving interests, which can be detrimental to the best interests of the project.

Disagreements among the stakeholders are to be expected, and the resolution thereof is part of the stakeholder management process. Conflicts within the project team, especially over objectives, can be fatal. It is almost impossible to hide these conflicts from key stakeholders.

SHIFTING LEADERSHIP

Managing Traditional Projects	Managing Nontraditional Projects
Most people believe that the project manager who started the project should finish the project. Continuity is essential unless the project gets into serious trouble.	The longer the project and the more influential the stakeholders, the greater the chance that there will be a frequent change in the leadership of the project. This could also be the result of a change in stakeholders.

On traditional projects, the leadership of the project can be expected to remain intact for the duration of the project. The same holds true for the governance. But on complex projects, leadership can change. The longer the project, the greater the likelihood that there will be a change in leadership.

Changes in the leadership style of the project, or stakeholder leadership, can prove to be detrimental to the project. Some project managers delegate significant authority to the project team for decision making, while other project managers may adopt a more authoritarian leadership style. Changes in leadership can also destroy the fervent belief, which may have taken years to develop.

During times of crisis, it is expected that a more authoritarian leadership style will appear on the project team and on the governance board. Not only is this normal, but, in most cases, necessary. However, having a change in leadership just for the sake of change can itself be detrimental. On long-term projects, people eventually develop "comfort zones" in how they will be treated and what expectations management has of them. Changes to the comfort zones can destroy morale.

WAGE AND SALARY INCONSISTENCIES

Managing Traditional Projects	Managing Nontraditional Projects
It is understood that some wage and salary inconsistencies may exist, and their impact on the project will be negligible.	Major problems can occur as a result of wage and salary inconsistencies. This can lead to employee turnover, especially among the employees with the key skills.

Most traditional projects use in-house resources or contractors. Complex projects, however, use partners and virtual teams. Problems can occur when these existing salary disparities are exposed. As an example, people who are direct reportees to the project manager may discover that some of their counterparts in their partner companies are earning higher salaries but may have lower credentials for the work they are doing.

A similar situation can occur when team members in partner companies discover that they are earning substantially less money than other team members and believe they are equally as qualified. In this case, exposing the salary differences could lead to a sabotage of the project.

Based on the way that project costs are captured and reported, earned value measurement can expose the salary differences, especially if the work performed on the project is charged back to the project as the actual salary, fully loaded, rather than a blended labor rate fully loaded. No matter how hard we try, salary disparities will eventually surface, and the project manager must be prepared to address this with project team members.

HIGH STAKES

Managing Traditional Projects	Managing Nontraditional Projects
Traditional projects generally do not have high stakes other than potential market share.	High stakes generally are accompanied by high pressure. If the pressure becomes excessive, people can be distracted from the real problems.

Traditional projects generally do not have high stakes other than perhaps a bonus for the project team and market share for the company. People do not necessarily expect promotions as a result of the outcome of a single traditional project on which they have worked.

Complex projects are viewed as career path opportunities as long as the project is completed successfully and bought into by the stakeholders. Unfortunately, high stakes are often accompanied by excessive pressure and a fervent belief to achieve. A fervent belief may take years to create but can be destroyed relatively quickly by excessive pressure on the project team.

Not all stakeholders may agree that the project is a high-stakes game. Generally, the higher the stakes, the more involved the key stakeholders. Sometimes, high stakes and excessive pressure can distract people from the real problems at hand. People may be willing to take excessive risks and approve perhaps unnecessary scope changes because of the high stakes. The good news, however, is that with high stakes you may find it easier to gain approval for additional funding for scope changes.

CULTURE

Managing Traditional Projects	Managing Nontraditional Projects
No cultural change is required. Culture is supported by a structured enterprise project management (EPM) system that requires linear thinking only.	Managing complexity may require a cultural change and nonlinear thinking. Complexity generally creates thinking— perhaps even out-of-the-box thinking.

The development of an enterprise project management (EPM) methodology is usually based on the existing or desired culture of the company. Every company has its own culture and a methodology designed to support that culture. On traditional projects, with most of the resources being assigned from within the company, the project manager generally has to deal with just one corporate culture.

Complex projects require nonlinear thinking and nonlinear applications of the project management methodology. Companies that have spent years creating an EPM system based on a linear thought process may discover that the culture cannot easily accept changes brought about through the execution or results of complex projects. People sometimes get so set in their ways that any disruption from their current comfort zones will cause them to rebel and revert to the cultural behaviors with which they feel most comfortable.

On complex projects, the project manager may discover that each partner has their own culture as does each virtual team. Some cultures foster cooperation, while others do not. Complex projects generally require creative thinking with out-of-the-box solutions. Some cultures do not allow for this and simply follow the straight-and-narrow path.

MULTIPLE CULTURES

Managing Traditional Projects	Managing Nontraditional Projects
One culture usually exists throughout the company, and the project's culture is compatible with the company's culture.	Large projects must endure multiple cultures, many of which are not compatible with the project manager's desired culture. The worst scenario is when the host's culture is not compatible with the project's culture.

As previously stated, traditional projects generally have the luxury of dealing with one, and only one, culture. Complex projects must endure multiple cultures, many of which may not be compatible with the culture the project manager desires. This is particularly true when working with developing countries. In some of these countries, even key stakeholders may not be allowed to make decisions without getting approvals from senior government officials. In this case, key stakeholders are just messengers. Another problem occurs when a crisis arises. The project manager may desire to have all problems brought to the surface immediately for resolution, while some partners may be afraid that the bearer of bad news will be beheaded.

Perhaps the most difficult situation is when the host's culture is not compatible with the project manager's culture. Although we often profess that the customer is always right, this might not be the case when two diverse cultures cannot work together. Some cultures focus on teamwork, whereas other cultures focus heavily on following policies and procedures. Some cultures focus on quality of work, while others may focus on just completing the deliverables. Religion can also come into play. Some cultures focus on cooperation under any situation, whereas other cultures focus on cooperation only if it comports with the "correct" religious or political practices.

MULTICULTURAL TEAMS

Managing Traditional Projects	Managing Nontraditional Projects
For these types of projects, there is generally one, and only one, culture to deal with when managing the team.	These types of projects generally require multicultural project leadership.

On traditional projects, there is generally only one culture to work with on the project. On complex projects, as stated previously, there will be more than one culture. There are two instances that could arise. In the first instance, the diverse cultures remain in the partners' companies or on the virtual teams where the resources are not under the direct, daily supervision and control of the project manager. In the second instance, the resources are physically removed from their company and placed under the direct control and supervision of the project manager.

In both cases, the project manager is required to have a multicultural leadership style. Working with people under your direct control may be more difficult than working with remote teams. You must understand the culture that these workers come from and how well or how difficult it will be for them to adapt to your leadership style and project environment. As an example, some cultures may have 20 or 30 paid holidays a year, whereas in the United States, there are typically 11. In some cultures heavily based on religious beliefs, team members may stop work to pray multiple times a day. This could affect the number of productive hours a day people will be available for work. Some cultures may have a low level of technology or tools available for the workers, and, if the workers are relatively set in their ways, they may not wish to learn how to do their job differently. In any event, the project manager must develop a multicultural leadership style to work with these people.

SHIFTING OF KEY PERSONNEL

Managing Traditional Projects	Managing Nontraditional Projects
Project managers prefer to have the same people assigned to the project from beginning to end. Project managers expect to lose key resources during a crisis, but it is usually kept at a minimum.	The shifting or loss of key personnel is usually beyond the control of the project manager. With a multitude of stakeholders, some will most certainly shift key resources for their own best interest rather than for the best interest of the project. The longer the project, the less likely it is that the project manager will be able to retain key personnel for the duration.

On traditional projects, we generally have the luxury of being able to retain the critical resources for the duration of the project, at least on a part-time basis. On longer-term projects, the chances of retaining the critical resources for the duration of the project are remote. Resources with critical skills will be in demand by more than one project.

Resources that are within partners' companies or assigned to virtual teams may be under the control of their stakeholders. It is possible that the stakeholders will shift the assignment of the critical resources based on their own self-interests rather than the best interest of the project. Shifting of resources at an inappropriate time could cause the project to endure a significant slippage.

A point should also be made about resource management processes—those processes used to train and select for, assign to, and remove human resources from projects. In organizations where no such processes exist, and because of the importance of those resources to project success, it's always more difficult to properly manage projects. The effects of this are even more exaggerated on the management of complex projects, reducing even further the probability of project success.

QUANTITY OF RESOURCES

Managing Traditional Projects	Managing Nontraditional Projects
Even though most companies are running "lean and mean," sufficient resources exist and are available for the project.	Vast resources may be required, and negotiation for the resources may be beyond the control of the project manager.

Most American-based companies are running "lean and mean" with regard to resources. On traditional projects, because of the relatively short time frame, project managers can generally squander up sufficient resources for the project. But on complex projects, there can be a significant difference between the resources needed and the resources available.

Vast resources are often required, and sometimes during the approval stage of the project, very little attention is given as to where the resources will come from and the quality of the resources available. Companies in some developing countries are under the mistaken belief that, if you partner with a company from an industrialized country, that company will have vast resources that can be assigned to the project. This is certainly not true.

The exact quantity of resources is generally not determined until planning begins. By that time, funding may have already been approved as well as the commitment of deliverables by a certain date. The project manager may then have to negotiate with the host and various stakeholders for the necessary resources. The negotiation process, and the ultimate control of the resources, may by that time be beyond the control of the project manager.

QUALITY OF THE RESOURCES

Managing Traditional Projects	Managing Nontraditional Projects
Project managers have some say in the quality of the resources required during project staffing activities. Functional managers may accept accountability for the quality of the assigned personnel.	Each stakeholder may assign resources based on availability, politics, and personal relationships rather than according to a required skill set. Project managers may have no input into project staffing activities.

When you manage a traditional project where most of the resources are assigned from an in-house pool, you normally know the difference between pay grades, such as a junior engineer, engineer, and senior engineer. However, on complex projects, with resources being assigned from partner companies and stakeholders, the quality and skill level of the resources can be quite different. For example, in one company, a junior engineer could be someone who graduated from a two-year technical school. An engineer could simply be someone with a college or university degree even though they have no technical experience. In industrialized countries, a senior engineer in one company could be someone with 10 or more years of experience, whereas in a company in the developing world, it could be someone with only 3 years of experience. Also, in some countries, the criteria for promotion to a higher pay grade may be based on the political party to which one belongs.

Project managers may have very little input into the staffing activities. Even if the project were provided with resumes of the resources to be assigned, it may still be almost impossible for the project manager to determine the exact skill level of the resources.

AVAILABILITY OF RESOURCES

Managing Traditional Projects	Managing Nontraditional Projects
Functional managers generally have manpower availability schedules in order to fit the required resource skill set to the project. If problems exist, they are known during project staffing.	Not all stakeholders care about or are in agreement with the skill sets needed for the project. Some stakeholders may not even know the skills of their own resources and simply assign whoever may be available at that time.

On traditional projects, the resources promised during the staffing process, assuming that the start of the project is relatively close to the staffing negotiations, will most likely be available. What is promised is normally received, including the skill set desired.

On complex projects, not all of the stakeholders will abide by their agreements on the type of support they will provide to the project. Providing head count may be more important than the quality of the resources. Also, the stakeholder assigning the resources may neither understand the quality of the resources needed nor the credentials of the people to be assigned. Simply stated, the stakeholders may have different agendas. Some stakeholders may believe that assigning poor resources is acceptable because they will get the necessary training while working on the project. No concern is given as to who will be doing the training of these people.

There is also the risk that stakeholders will not abide by agreements with other stakeholders. For example, one stakeholder may lose interest in the project and pull resources off of the project if his company is affected by an economic downturn. Another stakeholder may then pressure other stakeholders to pick up the slack.

CONTROL OF THE RESOURCES

Managing Traditional Projects	Managing Nontraditional Projects
Project managers have some degree of control of the resources, either directly or through the functional managers.	Project managers have very limited control of the assigned resources. Project managers may not be able to remove resources without political intervention.

On traditional projects, all resources except for subcontractors are directly under the control of the project manager or indirectly under his or her control through the line managers. On complex projects, the project manager may have limited control at best.

Control of the assigned resources is through the stakeholders or senior management in the companies from which the resources are assigned. Project managers may not be able to hire people and, likewise, may not be able to fire people. Project managers may also have virtually no input into the workers' performance reviews. The project manager may have no influence or clout in getting the resources assigned full time rather than part time. Consequently, the project manager may not be able motivate project team members through reward or punishment.

Project managers must work closely with key stakeholders to obtain resources with the required skills, and to retain them for the duration of the project if necessary. Without effective stakeholder management, and a multicultural leadership style, control of the resources may be impossible.

WORKER RETENTION

Managing Traditional Projects	Managing Nontraditional Projects
Worker retention is reasonably stable. The project manager can expect that the workers with the critical skills will remain on the project for its duration.	Worker retention may be a problem. Workers may not have any loyalty to the project or the company. Worker turnover, especially those with the most critical skills that are often in demand, is an important issue on long-term projects and impossible to control.

On traditional projects, the project manager normally works closely with functional managers for retention of the workers. If the project is an in-house effort, the line managers usually feel some degree of ownership and loyalty to the project, and worker retention is possible unless a higher-priority project occurs or some other crisis must be resolved.

On complex projects, worker retention can be a problem. Project managers usually work with the key stakeholders for resource assignments rather than the functional managers in each stakeholder's company. If the functional manager does not feel loyalty to the project, then the same feeling will permeate down to the worker.

Worker turnover is a critical issue and beyond the control of the project manager. It is highly unlikely that functional managers will assign their best people to a long-term project. If the resources are high quality and can make the functional manager look good in the eyes of his or her superior, then it is unlikely that the line manager will be willing to release these people for an extended period of time. The resources may be assigned initially to appease the key stakeholders and the project manager, and then some situation will arise causing the reassignment of those resources and their replacement by lower-grade resources, perhaps even some resources that the line managers might prefer to get rid of.

Chapter

7

PROCUREMENT
MANAGEMENT

MATERIAL/SERVICE REQUIREMENTS

Managing Traditional Projects	Managing Nontraditional Projects
The end product is well defined at the beginning of the project, making it easier to develop bills of material and to define required services.	The end product is not well defined, making it difficult to predict what will be needed against specific time frames, thereby necessitating nonstandard procurement arrangements through alliances or partnerships.

The Procurement Management knowledge area of the *PMBOK®️ Guide* deals with the processes around procuring materials and services external to those that are available from within the organization managing the project. Those processes include planning the required procurements, conducting the procurements (e.g., soliciting bids, selecting vendors, developing contracts), administering procurement contracts, and closing out those contracts at appropriate points within the project.

In traditional projects, the output (product, service, result) is normally able to be defined in some detail during (and even prior to) the project planning process, and the requirements associated with that output are also relatively well defined. With this being the case, the components and subcomponents that make up the end product can be easily identified during project planning. Despite changes to the end product that may occur during the course of the project and subsequent contractual changes with material and service providers, the procurement processes are normally fairly stable. They may be managed individually by the project team, or they may be managed by a procurement department representative on the project team. The associated procurement practices are usually governed by the organization's published procurement procedures.

Complex projects may pose a challenge to the typical procurement process. This is particularly apparent in research and development (R&D) and new product development projects, where it may be difficult to predefine the actual end product. In complex projects, it is not always possible to know what specific materials and services, and in what quantities, will be needed, and when they'll be needed by the project team. In such cases, standard organizational procurement procedures and arrangements may not suffice.

One approach that has worked on complex projects is that of developing partnerships or alliances with vendors and committing to arrangements in which both parties share the risks and benefits of successful project outcomes. At some point in time, particularly with the spread of complex projects, these new arrangements may become part of an organization's standard procurement practices.

BOT/ROT CONTRACTS

Managing Traditional Projects	Managing Nontraditional Projects
The seller develops the end product and quickly transitions it to the client/ end user with limited guarantees on long-term costs/use.	The increasing requirements by clients (owners) for build-operate-transfer (BOT) and refurbish-operate-transfer (ROT) contracts place a greater onus (risk) on the seller to guarantee the long-term costs and effective use of the end product. This significantly increases the duration and costs of the project as sellers go to greater lengths to protect themselves.

It's been general practice that the project team will develop a project product, train the buyers to use and maintain it, and then quickly transition the product, upon its acceptance, to the buyer. The buyer (the owner) normally would then be responsible for use and upkeep of the product and expect it to meet its use specifications for a period of time defined by the procurement contract between the buyer and the project organization (the seller).

In the past decade, a rarely used contract type has been gaining more use. It is known as a build-operate-transfer (BOT) or refurbish-operate-transfer (ROT) contract. To date, these contracts have been almost exclusively used for large infrastructure projects. In these contracts, the project organization is contracted not only to build/refurbish a facility, but also to operate it for a specified number of years and ensure its adherence to a predefined set of requirements over an extended period of time. The seller/operator receives the financial benefits for these operations and, at a point in time specified in the contract, the facilities are turned over to the buyer at no cost. An example of such an arrangement is the Panama Canal.

BOT/ROT contracts transfer much of the risk onto the selling (project) organization. The seller typically bears the cost of construction. They therefore must be assured that their operating revenues, once the project is built, will not only offset their costs, but also produce a decent overall profit during the time they operate the facilities they've built. This frees the owner, normally a government or government agency, from bearing the cost of building the facilities. And at the end of a specified period, the seller transfers the facilities to the owner. The owner may then decide to continue operations on their own, or they may contract with the seller to further continue operations for a period of time.

CONTROL OF VENDORS

Managing Traditional Projects	Managing Nontraditional Projects
The qualified vendors are generally all local, and the preferred communications medium is face-to-face meetings.	Vendors, particularly in cross-geographic projects, are spread far and wide, making them less accessible to, and increasing the difficulty of control for, the project manager. This distance also necessitates the need for the use of multiple communications media, the reliability of which in some parts of the globe makes control an even greater issue.

If one considers the communications required to interface with a vendor, they can be substantial. First, there are the communications around conducting procurements: identifying appropriate vendors; inviting them to bid on contracts; holding bidder conferences; interviewing and selecting the vendors; and negotiating their contracts. Next, there are the communications associated with project planning and execution: the vendors need to be included in the project kick-off activities; the project manager needs to get updates to project activities for which they're responsible; their work product needs to be reviewed and tested; and the project manager may need to take corrective action (possibly resulting in contract termination) during the course of the project. Finally, getting vendor feedback should be part of the project closing process.

In traditional projects, the preponderance of vendors are located geographically proximate to the project organization. This makes command and control far easier for the project manager, particularly when the projects are being performed in industrialized countries (communications media are generally more extensive and reliable than in the developing countries). When you consider the possible complexities of transnational (not to mention transcontinental) projects, the difficulties of interfacing with vendors become readily apparent.

The project manager needs to be well briefed on the availability and competence of potential vendors in the countries in which the project will take place. The project team must have some knowledge of the procurement methods that exist in each of those countries, as well as any potential ethical considerations that may have to be considered. While occasional face-to-face meetings may be possible, they will be fewer than ideal. The project team will have to fully understand the available communications media, their capabilities, and their shortfalls. And risk-response strategies associated with the shortfalls must be developed and implemented. Further complicating the picture are the differing judicial systems that would complicate transnational projects, with the attendant differences in contract law. A significant legal presence may be required as part of, or to be available to, the project management team.

REGULATIONS GOVERNING VENDOR SELECTION

Managing Traditional Projects	Managing Nontraditional Projects
While some regulatory influences exist on selection of vendors (e.g., minority and women-owned businesses), the procurement process is relatively straightforward and its rules are clearly understood by the seller, the owner, and potential subcontractors.	In projects that span multiple nations, the project management team must be aware of all of the rules and regulations governing the selection and use of vendors. If the project spans many political geographies, the complications that can arise may be staggering.

Many countries and localities have rules and regulations governing vendor selection and contracting. In the United States, for example, many states and localities have regulations encouraging the participation of minority and women-owned businesses (MWOB). Most nations have laws governing sales, both internal and external, of various types of products. Some products, like pharmaceuticals, high technology, and defense-related items, have stringent regulations associated with their sale. There are even stricter regulations around sales of these products to external buyers. And these rules and regulations vary from state to state and country to country.

Let's consider a project in which medical devices will be used. If these devices need to be purchased by project teams in various parts of the globe, different rules associated with supplier requirements will apply. The European Union requires that vendors of these devices be registered with an address in Europe. In Canada, vendors must obtain an establishment license. In the United States, these vendors must be registered with the Food and Drug Administration. In Japan, vendors are required to have one type of license for medical devices manufactured in-country, and another type of license for importing those devices.

The potential impact of these regulations can significantly affect the competing demands of a project—scope (more work to be done in vendor vetting and selection), time (longer vendor selection processes), cost (greater cost), quality (limited availability of quality vendors), risk (greater risk associated with work quality and reliability). In some companies that regularly do projects that involve all of these factors, people who are specialists in this area are frequently assigned to the project team to help understand the potential consequences of regulatory issues.

Finally, consider the consequences on the project if vendor selection is largely out of the control of the project management team. In highly politicized projects, and in much of the developing world, the project management may be told that they are limited in the selection of vendors to only those that are approved by various government ministries, regardless of the vendor's technical approach, technical capabilities, management approach, understanding of need, previous history, and so on. What alternatives does the project manager have when faced with such a situation?

IMPACT OF STAKEHOLDERS

Managing Traditional Projects	Managing Nontraditional Projects
Vendor selection processes are well defined, and decision making about vendor selection is limited to a small number of stakeholders and is, therefore, relatively quick.	The increased number of stakeholders spread over greater geographic distances and with competing agendas can make vendor selection a more onerous process.

As projects become larger and more complex, the number of key stakeholders may increase. This is readily seen in projects that involve partnerships or strategic alliances. In large, politically sensitive projects, key stakeholders may include cabinet ministers, their key staff members, and others from affected government agencies. Consequently, in projects that span multiple political entities, the number and types of potential management problems increase exponentially.

Consider the effects on project communications channels as the number of key stakeholders increases. As noted in Chapter 10, in a project that has 5 stakeholders, the number of communications channels is 10 ($n \times (n - 1) / 2$). Add just 10 more stakeholders and the number jumps to 105. That could amount to a communications nightmare for the project manager.

Think about what that might mean in a project that is directionally challenged and needs quick decisions made at significant milestones. Even assuming that all stakeholders have bought into the project objectives and have agendas that are aligned with those of the project management team, the difficulties associated with decision making in that environment are very challenging. Now consider that it's rare that alignment exists among all stakeholders, throw into the mix competing agendas for limited funds among governmental agencies, and we have a recipe for a cake that will not rise.

ADVERSARIAL PROCUREMENT POSITIONS

Managing Traditional Projects	Managing Nontraditional Projects
Sellers and buyers have different goals and adversarial procurement positions. The seller is trying to maximize profits while delivering the minimum acceptable end product. The buyer is trying to minimize cost for the best end product possible.	These adversarial procurement positions can lead to significant project risk in complex projects. New forms of procurement arrangements are needed to reduce the impact of these arrangements or change adversarial positions to collaborative ones.

In many instances, parties to project contracts have goals that are inharmonious. The buyer is trying to resolve a business problem or exploit a business opportunity in the least expensive manner, while the seller is trying maximize its profits with the least amount of work that fulfills its contractual obligations. These competing goals tend to set up adversarial positions between buyers and sellers at all levels (owner/ prime contractor, prime contractor/subcontractor) of the contract. On traditional projects, these risks are well known and can generally be dealt with using risk management processes. However, on complex projects, these risks can escalate precipitously, making the use of traditional risk processes sometimes unfeasible.

New agreements among all of the key stakeholders have evolved in which the owner, the contractor, and all subcontractors work under a voluntary collaborative working arrangement (CWA). The goal of this type of agreement is to minimize (if not eliminate) the adversarial postures of the parties, and to form a team that will share in both the benefits and the liabilities of the project outcomes. This requires a certain degree of trust among all of the parties, which is not always easily attainable.

The intended benefits of CWAs may include[1]:

- Improved predictability of project outcomes (time, scope, cost)

- Continuous process improvement

- Development of long-term relationships and efficient supply chains

- Minimizing risk

- Reducing costs

- Promotion of innovation among team members

- Acceptance of project management approach

- Ability to use traditional project management control/assessment tools (EVMS [Earned Value Management System])

---

[1] Adapted from K. Remington and J. Pollack. *Tools for Complex Projects*. Cornwall, UK: MPG Books Ltd., 2007, p. 104.

MULTIPLE CONTRACT TYPES

Managing Traditional Projects	Managing Nontraditional Projects
The buyer and seller agree on a clear vision for the output of the project, and they use a single contract type (e.g., time and materials, firm-fixed-price) for the entire project.	The output of the project or the direction to achieve that output may not be clear at the beginning of the project, necessitating the need for multiple contract types through the project. For example, when the output is not initially definable, a cost-plus contract is used until the output is clearly defined. From that point on, a firm-fixed-price contract can be used.

On projects that are done for external clients, the project management team needs to consider the type of contract that is used and its subsequent implications on risk sharing between the buyer and seller. Typically, a single contract type is chosen for the work of the entire project. The *PMBOK® Guide* suggests (on p. 321) that the firm-fixed-price (FFP) contract is the most frequently preferred (and in some cases, demanded) contract type. In FFP contracts, the preponderance of the risk falls on the shoulders of the seller (project organization). If the seller can deliver the end product within the contract's specified parameters and at or below the project budget, the project is considered to have been a success. Conversely, every dollar, euro, or yen that the project overruns its budget, the less of a profit the seller ends up making.

That may work fine on a project in which the end product's specifications are well defined by the time the contract is signed (when the price, timeline, and quality attributes associated with the end product are defined). That may not be the case in complex projects where the end product is only vaguely defined prior to project planning. How can a seller reasonably expect to have any assurances of making a profit on a project in which its remuneration is set prior to understanding the work that must be performed or the materials that might be needed? And how can we expect a seller to even bid on such work? Nonetheless, it happens as a regular occurrence.

In such cases, we may expect to see the use of mixed contract types over the period of a single project. The buyer and seller will agree to use some version of a cost-reimbursable contract to cover the part of the project, during which the project's end product is more clearly defined through progressive elaboration and prototyping. At that point in time, a second FFP contract is drawn up to cover the work in the remainder of the project. In some cases, the buyer may even invite bids from other vendors for the remainder of the project.

8

QUALITY MANAGEMENT

"SATISFICING" ZONES

Managing Traditional Projects	Managing Nontraditional Projects
Well-defined end products have specified quality attributes that must be met for the product to be acceptable.	At the beginning of the project, the end product may be only vaguely described, and stakeholders, at some point, agree on a quality spectrum within which the solution is "good enough" for their purposes.

One of the outputs of the Plan Quality process, described in Section 8.1.3.2 of the *PMBOK® Guide*—Fourth Edition, is quality metrics. "A quality metric is an operational definition that describes, in very specific terms, a project or product attribute and how the quality control process will measure it."[1] Traditional projects, with a clear definition of their outputs, are able to assign specific attributes to deliverables throughout the project and for the project end deliverable. Examples of these specific metrics range from one such as "a steel cylinder with a length of 15.5 centimeters and a diameter of 2.1 centimeters with tolerances of ±.05 millimeters for both dimensions" to an end product that has "an MTBF [Mean Time Between Failure] of 3,500 hours," depending on the deliverable whose attributes are being described.

As we have noted throughout this document, one of the uncertainties of complex projects might be an inability to specifically describe the end product in the initial project phases. The desired specificity may not be available until after a number of prototypes have been developed or a specific product has evolved through the research and development (R&D) process. So, instead of specific product attributes, project teams and clients may need to settle on general attribute ranges of "satisficing zones." The concept of "satisficing" was first developed by Herbert Simon, one of the pioneers in the field of artificial intelligence, in a book he wrote in 1957 on decision making and administrative behavior.[2] Simon talks about people trying to make decisions with information available in the present coupled with uncertainty about that information's accuracy in the future (when, let's say, an end product will be finally developed). Because of our inability to accurately predict future conditions, the rationality of these decisions will be bounded by the accuracy (or inaccuracy) of current information (he called this "bounded rationality"). So we need to make these decisions

[1] Project Management Institute, *A Guide to the Project Management Body of Knowledge*, 4th ed. Newtown Square, PA: Project Management Institute, 2008, p. 200.
[2] Herbert Simon, *Administrative Behavior*, 4th ed. New York: Free Press, 1997.

by "satisficing" (a combination of satisfying and sufficing)—settling for that which may not be an optimal outcome but which will be good enough for the task at hand. In fact, in some complex projects, as work unfolds, prototypes are modified, or discoveries are made, the target strictures around this satisficing zone may need to be modified in the face of current conditions (changed market or competitive conditions, new legislative regulations, new zoning laws, etc.).

DIFFERENT LIFE CYCLES

Managing Traditional Projects	Managing Nontraditional Projects
Traditional life cycles are used, providing gates at which project technical progress can be measured.	Different types of life cycles need to be used, depending on the specific technical and business needs of the project. All of these require continuous communications among the project team, end users, and other stakeholders to help develop quality attributes in a fluid environment.

Since life cycles are used to a good extent to ensure product quality, it is instructive to examine the suitability of specific types of life cycles to traditional versus complex projects. The *PMBOK® Guide—Fourth Edition*[3] cites three categories of project life cycles (phase-to-phase relationships):

- *Sequential*: in which a succeeding phase can only begin after its preceding phase is completed (a waterfall life cycle).

- *Overlapping*: essentially a sequential life cycle in which phases can overlap (e.g., a design/build project).

- *Iterative*: where only one phase is fully planned at any given time, and work on the next sequential phase is planned as work progresses on the current phase (e.g. a spiral life cycle).

The *PMBOK® Guide*—Fourth Edition goes on to say that it's possible that more than one category of life cycle may be used during a multi-phase project (e.g., using iterative for first two phases, sequential for last three phases).

In her book *Managing Complex Projects*, Kathleen Haas spends quite some time looking at the appropriate life cycles to use in a range of project complexities.[4] For independent (relatively traditional) projects, where the product requirements are well understood, project duration is short, and the project team members are competent and limited in numbers,[5] she cites four appropriate life cycles and circumstances under which each might be selected: *waterfall* (sequential), *modified waterfall* (overlapping), *rapid application development* (RAD), and *Vee* (for system component integration and verification). She also mentions Critical Path Project Management as an emerging practice.

---

[3] Project Management Institute, op. cit., pp. 21–22.
[4] Kathleen Haas, **Managing Complex Projects:** *A New Model*. Vienna, VA: Management Concepts, Inc., 2009, pp. 73–111.
[5] Please see Haas, *Managing Complex Projects*, for a complete description of the complexity models and their associated life cycles.

On moderately complex projects, those of a longer duration (three to six months), a larger number of team members (both internal and external), and an agreed scope subject to some change during the project, she discusses the *incremental delivery*, *spiral*, and *agile* approaches. All three approaches are iterative and allow for (and even encourage) scope change and redefinition, using end-user feedback and lessons learned to improve on the just-completed iteration. She also speaks about *Lean* and *Skunk Works* as emerging approaches for projects of moderate complexity.

Finally, for highly complex projects, those designed to bring about enterprise-wide change, with many areas of uncertainty (risk) and end products that are difficult to define, Haas notes two approaches that may prove feasible: *evolutionary prototyping*, and *eXtreme PM*. Both of these involve late design freezes, frequent contact with/feedback from the clients and end users, built-in redundancy, experimentation, and the exploration of multiple options done simultaneously. She cites some emerging approaches that seem to share many of these attributes.

The main thrust of all of this is that on complex projects, project teams shouldn't be limited to using life cycles that, while having worked on other traditional projects, may not provide utility (and may, in fact, impede progress) on their current project.

TECHNOLOGY

Managing Traditional Projects	Managing Nontraditional Projects
The project's technology is known, proven, and accepted by the project team.	The project's technology may need to be developed during the project. Users may need to be convinced of the use of the unproven technology.

During the course project execution, a wide range of technologies will be employed. In construction projects, these technologies may involve computer-aided drafting and design (CADD) systems for the design process, and building materials and machinery for the construction process. In software development projects, they may involve operating systems, middleware, applications, and the computers that run them. In traditional projects, these technologies have been used before by the project team members. They understand how to properly use these technologies and what their limitations are. The use of the technologies is accepted by team members. Should project team members using these technologies be unexpectedly unavailable, there are other available resources within the enterprise, knowledgeable about the technologies, who can replace them as needed. A complex project, however, may require the use of new or unproven technologies, and these technologies or their improper use may have a negative impact on the quality of the project's deliverables.

Take, for example, the use of a "new" building material. There is a well-known motivational videotape entitled *Four Hour House* (put out by the Building Industry Association of San Diego) that is meant to illustrate the principles of good project management and teamwork. The video shows a competition between two construction teams, populated by various construction trades, to see which team can build a complete house in the shortest amount of time. The winning team's time was just under four hours. To be able to complete the project in such a short amount of time, both teams used a relatively new, fast-setting concrete for their house's foundations, along with other prefabricated building parts. After the competition, these houses were subsequently sold and became the homes of several buyers. What is not mentioned on the video is that about 16 months after the homes were built, serious problems began to arise with their foundations due to the concrete that was used in their construction, and repairs were required.

The preceding is just one notable example of how new technology may affect quality. Consider the use of software applications used to determine the strength of a new aircraft's wings under various stress conditions. Let's say that you were aware that the management of the company building the aircraft had decided to use a new software application that involved the latest data on wing materials and was cheaper but more difficult to use than its previously used application. Would you be comfortable flying in that aircraft? Of course, product quality might not be the only victim of these circumstances. How do you think the project team members, whose professional reputations might be at stake, would feel about using an application, the results of which were, at least to them, unproven?

Considerable thought must go into the selection of, training on, and use of new, possibly unproven technologies. All risks associated with these technologies must be identified and thoroughly vetted, and alternative technologies (and the implications of their subsequent use) must be considered and factored into the project quality plan and all ancillary plans (e.g., risk, schedule, cost, procurement).

COST-BENEFIT ANALYSIS

Managing Traditional Projects	Managing Nontraditional Projects
The project's cost-benefit analysis is based on well-tested mathematical models.	New cost-benefits models may have to be developed, taking into account costs and benefits that may not easily be reduced to numerical formulas.

One of the tools and techniques in the *PMBOK® Guide's* Plan Quality process is the cost-benefit analysis. In section 8.1.2.1, it says "[t]he primary benefits of meeting quality requirements can include less rework, higher productivity, lower costs, and increased stakeholder satisfaction. A business case for each quality activity compare to the cost of the quality step to the expected benefit."[6] Interestingly enough, this is the only place in the *PMBOK® Guide* that speaks about cost-benefit analysis. When we speak about cost-benefit analysis, we want to understand the true value of the project to the organization. It's our contention that such a value is becoming more important in project management, as the need for regular project value analysis (both short and long term) becomes the norm.

The traditional cost-benefit analysis is based on well-tested mathematical models. Examples of these models include net present value, internal rate of return, and depreciation. These types of models, however, may be far too narrow in their scope. Measuring only tangible costs and benefits, as they do, overlooks other equally important factors like the benefits of the project's product, the reputation of the organization that produced the product, technological characteristics, and market value, benefits that may not easily be reduced to numbers.

Additionally, another related issue arises in projects whose complexity is associated with the parallel pursuit of multiple options (see the Different Life Cycles feature above) and end products that evolve over time (as in R&D projects). Not only do the cost-benefit evaluations need to be made prior to project initiation, but they also need to be done regularly at the appropriate phase gates to ensure that the evolving options, with their ever-evolving sets of features and functions, continue to provide the benefits that drove the project's initiation and meet the consequent changes in perspective on those benefits as the end product's definition is clarified.

---

[6] Project Management Institute, op. cit., p. 194.

NEW QUALITY BOUNDARIES

Managing Traditional Projects	Managing Nontraditional Projects
The project's output and associated quality requirements and measurement methods are predefined.	New end products and the use of new technologies require the development of new quality boundaries and measurement methods.

In project management practice, project deliverables (components, subcomponents, work package artifacts) all have acceptance criteria against which they are measured to ensure adherence to quality requirements. These quality criteria and the methods to measure the deliverables to ensure adherence to these criteria are defined well before project execution commences. The criteria are based on a history of quality standards associated with various technologies, materials, processes, and industries. For traditional projects these criteria are deemed appropriate (and are sometimes mandated by regulatory requirements), and the measurement methodologies are developed and accepted by all stakeholders.

In projects whose complexities are, at least in part, caused by the use of new technologies or involve the development of new products or substances, there may be a paucity or absence of appropriate quality criteria, and the quality measurement methodologies may need to be developed. Additionally, as a result of using the product developed by the project team, wholly unanticipated problems might arise, prompting the need for incremental quality criteria.

Let's consider the example of a project team of environmental engineers whose task is to employ an appropriate method of odor control in a project to build a new municipal wastewater treatment plant. Quality requirements for odor control in wastewater treatment facilities are well defined by regulatory bodies. And in traditional projects, these requirements would suffice. But let's say that during design of the new plant it was determined that current technologies would not produce the quality result required by those regulations. An attempt to either introduce or develop a new method for odor control might become one of the project team's goals. And if the new method that's finally employed ultimately produced pollutants that had hitherto not been detected (or even seen before), a new quality constraint might need to be developed and applied to the method just developed. This, quite naturally, would introduce delays into the project schedule and more than likely have an adverse effect on the project budget as well. So risks associated with product quality need to be well considered in setting stakeholder expectations around technology, scope, schedule, and cost.

Chapter

9

RISK
MANAGEMENT

COMPLEXITY, UNCERTAINTY, AND RISK

- Project complexity results from increased uncertainty about:

  - Stakeholders and stakeholder interactions

  - Duration

  - Funding

  - Scope changes

  - Resources

  - Potential risk interactions

- Managing complex projects is all about managing project risks and their possible interactions.

If we stop for a moment to consider what really differentiates traditional projects from complex projects, what becomes fairly obvious rather quickly is the significant increase in risk associated with complexity. If we look at either of the models on project complexity cited in this book (Haas, Remington/Pollack), each bases complexity on the degree of uncertainty (risk) associated with various project attributes. The Remington/Pollack model discusses risk associated with four project dimensions[1]: technical, directional, structural, and temporal. In the Haas model,[2] we assess the uncertainty associated with:

■ Cost/duration

■ Team composition and performance

■ Urgency/feasibility

■ Problem solution clarity

■ Requirements volatility/information technology (IT) complexity

■ Political sensitivity/multiple stakeholders

■ Organizational/commercial change

■ Risk, external constraints/dependencies

A successful project manager of complex projects is one who is able to live with and manage a good deal of project uncertainty or risk. In the next few pages, we will examine the *PMBOK® Guide's* risk management processes and see how, if at all, they are affected by project complexity.

---

[1]K. Remington and J. Pollack. *Tools for Complex Projects*. Cornwall, UK: MPG Books Ltd., 2007, p. 6.
[2]Kathleen Haas, **Managing Complex Projects:** *A New Model*. Vienna, VA: Management Concepts, Inc., 2009, p. 48.

RISK MANAGEMENT

Managing Traditional Projects	Managing Nontraditional Projects
Most project managers have a knowledge of risk management and perform it throughout the life cycle of the project.	Effective risk management may not be recognized as important and therefore downplayed. The identification of risks could lead to government interference and possibly project termination.

Traditional projects have a standard risk process that's used to identify, analyze, and manage foreseeable risks. Additionally, unforeseeable risk may be dealt with by using buffers for both cost (management reserve) and time (critical and near critical path buffers, particularly if using critical chain methodology). However, even in some traditional projects, it's astonishing how many project managers are afraid to even raise the issue of risk associated with their projects. The fear of discussing risk stems primarily from the potential of its leading to premature project termination. After all, the reasoning goes, why should an owner want to throw money into a project if there are lots of uncertainties about its being a success?

With their higher levels of uncertainty, this ambivalence about discussing risk attends complex projects to an even greater degree. And when you throw bureaucratic and political interference into the mix, particularly in government-funded projects (which many complex projects are), that fear of discussions around risk may eliminate it ever being constructively discussed.

Are those fears well founded? Perhaps in a few cases they are. But the work of most projects, both traditional and complex, needs to be accomplished regardless of the risks involved, thus minimizing the probability of termination. It is beneficial for all stakeholders to identify and understand the risks associated with their projects so adequate risk response strategies can be developed and implemented that will minimize potential adverse effects on the project's success. And there may be positive risks (uncertainties) that, if properly exploited, can increase the probability of positive project outcomes. In almost all instances, clients appreciate the discussion around project risks as it better enables them to factor it into their portfolio's exposure, and they can better understand how their involvement might help to properly respond to project risks, either positive or negative.

Finally, if a project is so very risk that its probability of success is very limited or it exposes the organization to a great liability, it's better for that to be known at the outset of the project so that it can be properly dealt with—through termination or other means.

IDENTIFY RISKS

Managing Traditional Projects	Managing Nontraditional Projects
The project manager, with the help of the sponsor, generally understands most of the risks, with technical risks usually taking center stage.	It may be impossible for the project manager to understand all of the risks affecting each stakeholder. Risks that most project managers do not get involved with, such as political or cultural risks, may be at the top of the list.

A project manager's ability to predict foreseeable risks is associated with his having encountered that same risk, or a similar risk, on a previous project at some point in his career. To some extent, project managers can learn about risks from those that other project managers have encountered, through lessons learned debriefings or discussions in the organization's knowledge management system (assuming one exists), risk management check lists, and risk breakdown structures (RBSs). But, by and large, risks are identified through personal experience.

Complex projects present the project manager with a greater array of potential risks than those normally encountered in traditional projects. And frequently, many of those risks emanate from areas with which most project managers have little experience. Consider, for example, that most project managers rise through the technical ranks. Their expertise is generally restricted to the technical area in which they have been trained and in which they have worked. As projects become more complex, different types of technologies are involved—many in which the project manager is not well versed. So her ability to identify risks in those areas with which she's unfamiliar is limited.

Now let's complicate that equation by making the project one in which a consortium of companies and/or governmental agencies from around the globe are involved. That adds multiple levels of uncertainty to the project in the forms of an increased number of unfamiliar stakeholders, multiple cultures and customs with which to deal, and political implications that may be out of the realm of the project manager's previous experiences.

The profile of a project manager who will be successful in managing complex projects is clearly different from that of a traditional project manager and must be tailored to the complexities of the project at hand. Additionally, the project management team needs to be composed of people whose skills and experiences both complement and supplement those of the project manager, making team member selection more difficult than normal as well (see Chapter 6).

UNEQUAL CONTINGENCY PLANNING

Managing Traditional Projects	Managing Nontraditional Projects
Contingency planning is performed by the project team and approved by the stakeholders. All of the players understand their role in contingency planning.	Not all of the stakeholders are equal in ability when it comes to contingency planning. Some may want to be actively involved, whereas others may prefer to delegate the responsibility.

As noted in Chapter 10, in traditional projects the number of key stakeholders is generally limited, and most understand their roles and participate in contingency planning. The large number of these stakeholders in complex projects reduces the probability that this will be the normal course of events.

The stakeholder group will consist of people of varying abilities and desired involvement in the risk planning process. The project manager must first find out which among them have an interest in being involved, which do not, and which want to delegate the responsibility to members of their staffs. This normally is a function of the "Identify Stakeholders" process in the Communications Management knowledge area, which was discussed earlier in the book. In getting this information, the project manager needs to be instructive in explaining to the stakeholders the purpose and workings of the risk management process. Doing so is important since a good portion of the stakeholders may never have been involved in, or have knowledge of, formal project management methods. Not doing so may result in too few or too many stakeholders in the process.

Once it's known which of the stakeholders want to be involved in contingency planning, the project manager needs to understand their skills, technical knowledge, and risk process knowledge. For those who have little knowledge of the process, some training might be required. It's essential for the project manager to get stakeholder buy-in to the risk management process being used and to understand how the stakeholders' political associations may play a role in their input to that process.

RISK ANALYSIS

Managing Traditional Projects	Managing Nontraditional Projects
Most foreseeable risks are independent of one another and have few, if any, interactions.	The probability and impact of risks are interrelated. The materialization of some risks could compound the impact of others. The impact of these interrelations cannot be depicted using standard analysis tools and techniques.

When we examine risks in traditional projects, we look at them as individual entities with their own probabilities of occurrence and potential impact if they materialize at some point in time during the project. We may occasionally see a connection between two or three risks, but they are rare and easily foreseeable. Not so in complex projects. Many references on complex projects view them as complex adaptive systems. Such systems frequently display attributes that emerge during the course of the project but are unforeseeable at the beginning of the project. These emergent attributes may include various forms of project risk.

One reason that it may be difficult to identify an emergent form of risk is that the project management team typically has itself buried in the day-to-day details of performing the project's work. To use a well-worn cliché, they can't see the forest for the trees. This is particularly onerous because emergent risk forms usually take on a life of their own, leading to yet more uncertainty in the project. If not detected early on, they can take the project manager down futile, sometimes irrecoverable project paths.

During the course of a project's normal execution, the project manager regularly reviews the project risk register with the project team. Each individual risk is reviewed for changes in probability of occurrence and impact on the project. New individual risks are identified as changes to the project occur, and they are placed in the risk register's severity hierarchy to enable the appropriate risk-response strategy and activities to be developed. That may not be sufficient for a complex project. Instead, the project management team needs to change its perspective. It needs to look at the items on the risk register holistically, looking for linkages and trends that would not otherwise have been noticed. In that way, they should be able to spot and address these trends when they start to occur and not when they become too difficult to sort out.

MULTIPLE OPTIONS ANALYSIS

Managing Traditional Projects	Managing Nontraditional Projects
A linear path from project start to end is feasible, and the attendant risks are associated with work on that path.	There may be multiple, emergent, or circular paths to reach the project's objectives and each set of options must be analyzed for its attendant risks.

In any project of over 20 or more activities, we look for ways of focusing on the most important activities and moving concerns around the others to the periphery. One way of determining which activities are important is to look at those on the project's critical or near-critical paths. Any slippage that might occur on any of those activities will result in a delay of the project. Once identified, we focus on ensuring that risks associated with these activities have appropriate risk strategies and action plans in place (or built into the schedule).

The difference with many complex projects is that there may not be a linear critical path on which to focus. In fact, particularly in projects where there are multiple options available for reaching the project's end point, there may be circular, multiple, or emergent (remember, these are complex adaptive systems) paths, each having its own set of activities and attendant risks.

That raises two concerns on the part of the project manager. First, how does one schedule a project that may be nonlinear (precedence diagram method—the most commonly used scheduling technique on which almost all scheduling software is based—is a linear tool)? And second, is there any way of focusing on specific problem areas with multiple critical paths?

To the first issue, there is a project networking technique—Graphical Evaluation and Review Technique (GERT), developed in the mid-1960s—that allowed for multiple project paths, each having associated probabilities assigned. While this specific technique has largely fallen out of use (reference to it was removed in the third edition of the *PMBOK® Guide*), the underlying concept of it—multiple probabilistic paths—is incorporated into the risk management software applications mentioned elsewhere in this section. Their use can be of significant value to project managers whose project schedules may not be plotted in a linear fashion.

As to the second issue, each of the multiple paths must be evaluated independently and all inherent risks evaluated as well. That is the only way that the stakeholders can understand the complete exposure of the project. Finally, if new paths emerge during the course of the project, a replanning effort for the schedule would be required.

RISK PRIORITIZATION

Managing Traditional Projects	Managing Nontraditional Projects
Risks are easily prioritized and segmented using the probability/impact (PI) analysis.	More comprehensive risk management tools for sensitivity and outcome confidence analyses are needed (e.g., @RISK®, Primavera Risk Analysis®).

Once project risk issues are identified in traditional projects, the process calls for them to have their severity analyzed through qualitative and, if necessary, quantitative methods. The risks are then prioritized based on the outcome of those analyses according to their severity score (normally calculated with the equation *probability X impact*) to the project. They are separated into three categories—high, medium, and low—using thresholds for each level. Each category of risks is addressed as follows:

- *High risks*. Risk-response strategies assigned and action plans to implement the strategies are developed. The activities in those action plans are then included in the work breakdown structure and consequently their cost is reflected in the budget and time in the schedule. This incremental time and cost are included in the contingency reserve.

- *Medium risks*. As with high risks, response strategies and action plans are developed. However, these action plans are filed and regularly reviewed throughout the project. They are implemented only if the risk materializes.

- *Low risks*. Because of their low probability and impact, the resolution of these risks, should they materialize, is left to the project team.

The problem with this approach in complex projects is that a *Probability X Impact* analysis may not be sufficient. It looks only at the individual risk issue and may not be able to identify the true significance of the risk in the context of the overall project workflow and linkages with the effects of other risks. To understand the true prioritization of the risks identified on the risk register, it may be necessary to implement other analytical methods. One such tool is called a Tornado diagram (shown in the slide on the opposite page). It is an outcrop of Monte Carlo simulations and can be done with several available software applications (e.g., @RISK, the risk module in P6, and PERT Master). The use of these tools will enable the project management team to develop a much more useful prioritization of its project's risks.

DETERMINING RISK RESPONSE STRATEGIES

Managing Traditional Projects	Managing Nontraditional Projects
The project management team selects the management strategy it deems most appropriate to the identified risks.	The large numbers of stakeholders will make developing a consensus on response strategy selection more difficult and time consuming.

In the normal progression of the risk management process, once a risk has been identified and its severity score has been determined through qualitative analysis, the project management team selects a management strategy appropriate to the risk and the options available to address it. For negative risks, these approaches can include avoiding, mitigating, or transferring, and for positive risks, exploiting, enhancing, and sharing. If nothing can be done about the risk or its impact is almost inconsequential to the project, acceptance is an appropriate strategy for both positive and negative risks.

In complex projects, the selection of risk-response strategy may not be solely in the hands of the project management team. Indeed, in a project with a number of key stakeholders, each of whom have significant interest in the project, those stakeholders may have a significant impact on how a risk's response strategy is selected and the action plan that is put into place to address the risk.

Let's say you're managing a project in which you've got partnership arrangements with two key vendors. A significant risk materializes, and each vendor proposes a different response strategy—options that will financially benefit each of them individually to the exclusion of the other. It will clearly take more time to do a thorough analysis of each option to justify selecting one over the other and to smooth out the feathers of the vendor whose option wasn't selected. Suppose now that you're managing that same project in a developing country, and the vendor whose option wasn't selected is a blood relative of a cabinet minister, who is also a key stakeholder in the project. The impediments to selecting the correct alternative and getting the funding released to implement it might be overwhelming.

MONITORING AND CONTROLLING RISK

Managing Traditional Projects	Managing Nontraditional Projects
Risks can be monitored on a regular (biweekly, monthly) basis with lessons learned aggregated and distributed at the end of a phase or at the end of a project	Risks must be monitored on a more frequent basis, with lessons learned aggregated and distributed to the project team as they occur.

After the start of project execution, project managers and their project teams monitor and control risk on a regular basis. Depending on the project's total duration, this is normally done on a biweekly or monthly basis. In the process of doing so, new risks that have been identified are added to the risk register, and they go through the same analysis/severity ranking/response strategy steps discussed earlier in this chapter. Doing so enables the project team to anticipate and proactively manage the risk events and not have the risk events manage them reactively.

Proactive risk management may not always be possible in complex projects. There may be so many uncertainties associated with the technical approach or with the end product that some level of reactive risk management must be borne by the project team. The project management team can approach this by assessing the areas of the project in which these risks are most likely to occur and developing quick-response teams to address them as they materialize (similar to military or police quick-response teams)—to be in a sense "proactively reactive." Naturally, the estimated costs of this firefighting must be built into the budget and incremental time for this has to be added into the schedule. Additionally, the project schedule must account for the limited availability of these team members to work on their scheduled project activities when they are occupied with their quick-response team duties. And to stay on top of emerging risks, the frequency of risk monitoring will increase to a weekly, and in some instances daily, basis.

Normal project management practice also calls for the development of lessons learned by project team members, not to mention their appropriate aggregation and distribution throughout the entire project team and to the rest of the enterprise. This is typically done at the end of each project phase and at the end of the project. However, the learning environment in complex projects is more demanding than in traditional projects, and the need for rapid collection, aggregation, and distribution of lessons learned is paramount. These activities must also be assigned to specific resources within the project team with their impact on project cost and schedule activities incorporated into the project plan.

TECHNICAL RISKS

Managing Traditional Projects	Managing Nontraditional Projects
The application of a formal project life cycle is generally not necessary.	To minimize technical risk, the project should have a formal phase-gate management process.

The second chapter of the *PMBOK® Guide*—Fourth Edition, discusses project life cycles, their characteristics, and their employment in project management. In short, a life cycle is a series of phases that separate a project into more easily manageable and measureable segments. The project life-cycle process is also called a "phase-gate" process because there is a virtual gate when moving from one phase to the next subsequent phase. The gates between phases are project assessment points, and in order for the project to proceed to the next successive phase, it and its deliverables must successfully negotiate the phase-end assessment. Among the purpose of these assessments is the need to ensure that the project's technical progress is proceeding at an acceptable pace and that the end deliverable is achievable within defined quality parameters. These are both typically determined at the beginning of the project by those who have initiated it, and may need to be modified during the project depending on project environmental changes (market conditions, competitive and industry pressures, etc.). So, in part, the use of the project life cycle serves to minimize technical risk.

Not all projects have the need to impose a phase-gate process. In fact, in many small and medium-sized projects there is no predefined life cycle. The more complex the project gets, the greater the need for a life cycle. The Remington/Pollack book discusses criticality of phases associated with the project complexity types. It says, for example, that "[t]he critical project phases for technical complexity tend to be the initiation and design/development phases."[3] And in her book, Kathleen Haas has three chapters devoted to appropriate life cycles for her three types of project complexity (independent, medium, and high).[4]

The use of a clearly defined life cycle with well-understood and appropriate measurements for each phase gate will go a long way in helping the project management team minimize technical risks as the project progresses.

---

[3] Remington, and Pollack, J., op. cit., p. 43.
[4] Hass, op. cit. pp. 79–111.

MANAGEMENT RESERVE

Managing Traditional Projects	Managing Nontraditional Projects
Typically, a nominal amount is set aside for managing unforeseeable (unknown) risks, depending on the industry/technology involved. Those funds are readily accessible, allowing for fast response to unforeseeable risks that materialize.	A significant amount of funds needs to be set aside for managing unforeseeable risks, whose probability of occurrence and impact on scope, schedule, cost, and quality far exceeds that of a traditional project. Due to the increased complexity of decision making, it will be more difficult/take longer to get access to reserve funds once the risks materialize, making it more difficult to respond quickly to unforeseeable risks.

E very project has both foreseeable and unforeseeable risks. High-severity foreseeable risks were addressed earlier in this chapter, and the funds for the activities associated with them are part of the project's contingency reserve. For medium- and low-severity risks, and for unforeseeable risks as well, funds are generally set aside in the project's management reserve. Management reserve is calculated as a percentage of the performance management baseline, typically between 5 percent and 20 percent, and its access is outside of the control of the project manager. Unlike the funds in contingency reserve, they are accessible to the project manager only after they are requested from and released by management.

In traditional projects, deciding on the amount to set aside for management reserve is relatively straightforward. Knowledge gained from previous similar projects instructs the project team and management in determining needed funds. Additionally, when an unforeseen risk does materialize, the project manager has relatively quick access to those funds with few levels of management approval needed. This access enables an equally quick response on the part of the project team to minimize the impact of the risk on the project. In complex projects, neither of these may be quite so easy to achieve. One potential problem for both issues lies in the number of key stakeholders involved and the levels of management, both of which tend to hinder the processes involved.

Arriving at a consensus on how much to set aside for the management reserve in a project that has no well-defined end deliverable and for which the technology being used in the project will evolve through acts of discovery can be daunting. The incremental layers of approval, coupled with the need for buy-in from a greater number of key stakeholders, will slow down this process even further. Further complicating matters is the fact that, due to the uncertainties involved in complex project, the level of reserve funding will necessarily be higher than normal. Even after such a consensus has been reached and the funds have been set aside, the multiple levels of approval for their release will erect obstacles that will slow down the team's ability to respond to unforeseeable risk, thereby causing the risk to have a greater impact on the project than would normally result.

COMMUNICATIONS
MANAGEMENT

STAKEHOLDERS

- Stakeholders are individuals or groups that may have a direct or indirect interest in the project and can be affected by the deliverables or ultimate value.

- Stakeholder management is the process of managing the expectations of the stakeholders without sacrificing your company's mission or vision.

Stakeholders are, in one way or another, individuals, companies or organizations that may be affected by the outcome of the project or the way in which the project is managed. Stakeholders can be affected throughout the project either directly or indirectly, or may function simply as observers. Stakeholders can shift from a passive behavior to becoming an active member of the team and participate in critical decisions.

On small or traditional projects, we generally interface with just the project sponsor as the primary stakeholder, and the sponsor usually is assigned from the organization that funds the project. This is true for both internal and external projects. But the larger the project, the greater the number of stakeholders you must interface with. The situation becomes even more potentially problematic if you have a large number of stakeholders, geographically dispersed, all at different levels of management in their respective hierarchy, each with a different level of authority, and language and cultural differences. Trying to interface with all of these people on a regular basis, especially on a large, complex project is very time consuming.

One of the complexities of stakeholder management is figuring out how to do all of this without sacrificing your company's long-term mission or vision. Also, your company may have long-term objectives in mind for this project, and those objectives may not necessarily be aligned to the project's objectives or each stakeholder's objectives. Lining up all of the stakeholders in a row and getting them to uniformly agree to all decisions is more wishful thinking than reality. You may discover that it is impossible to get all of the stakeholders to agree, and you must simply hope to placate as many as possible at a given point of time.

STAKEHOLDER COMMITMENT

■ Stakeholder management is driven by commitment. But stakeholder commitment cannot be obtained unless the stakeholders are sold on the value they will receive at the end of the project.

Stakeholder management cannot work effectively without commitments from all of the stakeholders. Obtaining these commitments can be difficult if the stakeholders cannot see what's in it for them at the completion of the project, namely, the value that they expect. The problem is that what one stakeholder perceives as value, another stakeholder may have a completely different perception or a desire for a different form of value. For example, one stakeholder could view the project as a symbol of prestige. Another stakeholder could perceive the value as simply keeping their people employed. A third stakeholder could see value in the final deliverables of the project and the inherent quality in it. And a fourth stakeholder could see the project as an opportunity for future work with particular clients.

Defining ultimate value success on a project has never been an easy task. Today, we believe that there are four cornerstones for success, and each can be seen as value as well:

- *Internal success or value.* The ability to have a continuous stream of successfully managed projects using an enterprise project management (EPM) methodology and that continuous improvement occurs on a regular basis.

- *Financial success or value.* The ability to create a long-term revenue stream that satisfied the financial needs of the stakeholders.

- *Future successor value.* The ability to produce a stream of deliverables that will support the future existence of the firm.

- *Customer-related success or value.* The ability to satisfy the needs of the customers or stakeholders over and over again to the point where you receive repeat business and the customers treat you as though you are a partner rather than a contractor or supplier.

GETTING STAKEHOLDER AGREEMENTS

It is essential that stakeholder agreements be reached as early as possible in the project, perhaps before the project actually kicks off. Failure to do so can lead to devastating results.

Getting stakeholders to be committed to the project and seeing the ultimate value in supporting it is easy as long as you are willing to allow each stakeholder to have their own views on the project, its objectives, success criteria, and ultimate value. But getting all of the stakeholders to reach an agreement is very difficult, if not impossible, especially if there are several stakeholders.

Another form of agreement involves developing a consensus on how stakeholders will interact with each other. It may be necessary for certain stakeholders to interact with one another and support one another with regard to sharing resources, providing financial support in a timely manner, and the sharing of intellectual property.

While all stakeholders recognize the necessity for these agreements, they can be impacted by politics, economic conditions, and other enterprise environmental factors that may be beyond the control of the project manager. Certain countries may not be willing to work with other countries because of culture, religion, views on human rights, and other such factors.

For the project manager, obtaining these agreements right at the beginning of the project is essential. Some project managers are fortunate in being able to do this while others are not. Leadership changes in certain governments may make it difficult to enforce these agreements on complex projects.

STAKEHOLDER ISSUES AND CHALLENGES

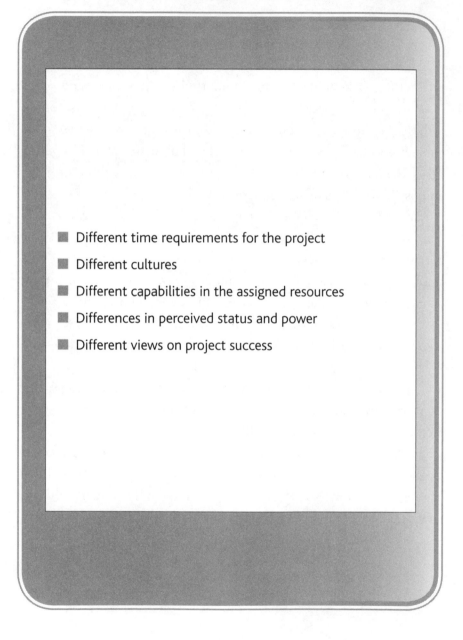

- Different time requirements for the project
- Different cultures
- Different capabilities in the assigned resources
- Differences in perceived status and power
- Different views on project success

It is important for the project manager to fully understand the issues and challenges facing each of the stakeholders. Although it may seem unrealistic, some stakeholders can have different views on the time requirements of the project. In some developing nations, the construction of a new hospital in a highly populated area may drive the commitment for the project even though the project could be late by a year or longer. People just want to know that it will eventually be built.

In some cultures, workers cannot be fired. Because they believe they have job security, it may be impossible to get them to work faster or better. In some countries, there may be as many as 50 paid holidays for the workers, and this can have an impact the project manager's schedule.

Not all workers in each country have the same skill level, even though they have the same title. For example, a senior engineer in one country may be perceived as having the same skills as a lower-grade engineer in another country. In some locations that may have a shortage of labor, workers are assigned to tasks based on availability rather than capability.

In some countries, power and authority, as well as belonging to the right political party, are symbols of prestige. People in these positions may not view the project manager as their equal and may direct all of their communications to the project sponsor. In this case, it is possible that the salary is less important than relative power and authority.

MAKING BAD ASSUMPTIONS

■ Do not assume that all stakeholders want the project to succeed. Their true feelings may not become apparent until the project is near completion. Reasons for not wanting success may include:

- ■ Team will be disbanded at the end of the project, and stakeholders may be unsure about their next assignment.

- ■ Loss of power and/or authority when project ends.

- ■ Loss of employment at the end of the project.

- ■ Radical change in the corporate culture if project succeeds.

- ■ Possible radical change in ongoing business processes.

- ■ A fear of having to learn new systems at completion.

- ■ An increase in pressure to use the new system.

It is important to realize that not all of the stakeholders may want the project to be successful. This will happen if stakeholders believe that they may lose power, authority, hierarchical positions in their company, or, in a worse case, even lose their job. Sometimes these stakeholders will either remain silent or even be supporters of the project until the end date approaches. If the project is regarded as unsuccessful, these stakeholders may respond by saying, "I told you so." If it appears that the project may be a success, these stakeholders may suddenly transform from supporters or the silent majority to adversaries.

It is very difficult to identify these people. These people can hide their true feelings and be reluctant to share information. There are often no telltale or early warnings signs that indicate their true belief in the project. However, if the stakeholders are reluctant to approve scope changes, provide additional investment, or assign highly qualified resources, this could be an indication that they may have lost confidence in the project.

ANOTHER BAD ASSUMPTION

■ Do not assume that key stakeholders understand their role and relationship with the project manager.

■ Some key stakeholders may desire to micromanage the project and can do more harm than good by usurping the authority of the project manager.

Not all stakeholders understand project management. Not all stakeholders understand the role of a project sponsor. And not all stakeholders understand how to interface with a project or the project manager even though they readily accept and support the project and its mission. Simply stated, the majority of the stakeholders are never trained in how to properly function as a stakeholder. Unfortunately, this cannot be detected early on but will become apparent as the project progresses.

Some stakeholders may be under the impression that they are merely observers and need not participate in decision making or authorization of scope changes. For some stakeholders who desire to be just observers, this could be a rude awakening. Some will accept the new role while others will not. Those that do not accept the new role usually are fearful that participating in a decision that turns out to be wrong can be the end of their political career.

Some stakeholders view their role as that of a micromanager often usurping the authority of the project manager by making decisions that they may not necessarily be authorized to make, at least not alone. Stakeholders that attempt to micromanage can do significantly more harm to the project than stakeholders that remain as observers.

It may be a good idea for the project manager to prepare a list of expectations that he or she has of the stakeholders. This is essential, even though stakeholders support the existence of the project. Role clarification for stakeholders should be accomplished early on the same way that the project manager provides role clarification for the team members at the initial kickoff meeting for the project.

VALUE CREATION

- Stakeholders view projects as value creation. Project managers should have the same view.

The fact that the deliverable is provided according to a set of constraints is no guarantee that the client will perceive value in the deliverable. It is true that clients track budgets and schedules, but it is the value at the end that makes the project a success or failure.

The ultimate objective of all projects should be to produce a deliverable that meets expectations and achieves the desired value. This should be the goal of the project manager as well as the client. While we always seem to emphasize the importance of the triple constraint when defining the project, we spend very little time in defining the value characteristics that we expect in the final deliverable.

The value component or definition must be a joint agreement between the customer and the contractor (buyer/seller) during the initiation stage of the project. Also, in the ideal situation, **the definition of value is aligned with the strategic objectives of both the stakeholders and the project manager.**

Warren Buffett emphasized the difference between price and perceived value when he stated, "Price is what you pay. Value is what you get." Most people believe that customers pay for deliverables. This is not necessarily true. Customers pay for the value they expect to receive from the deliverable. If the deliverable has not achieved value or has limited value, the result is a dissatisfied customer.

Some people believe that a customer's primary interest is in the quality of the deliverable. In other words, quality comes first! While that may seem to be true on the surface, the customer generally does not expect to pay an extraordinary amount of money just for high quality. Quality is just one component in the value equation. Value is significantly more than just quality.

STAKEHOLDER MANAGEMENT RESPONSIBILITY

- The project manager has the ultimate responsibility for stakeholder management even though others, such as the project sponsor or project management office, may participate.

There is a common belief in project management that whoever makes the decisions on the project has the ultimate responsibility for the project's success or failure. This is not true. While project managers may discover that a multitude of decisions are being made by the client and the key stakeholders, the ultimate responsibility for success or failure rests with the project manager. This is just like quality; project managers can and do delegate quality control work to others on the project team, including some quality-associated decision making, but the project manager retains the ultimate responsibility for the quality of the project.

It is not uncommon on large, complex projects that the role of the project manager becomes more of a facilitator or coordinator of decisions made by others. This happens because the project manager most likely has an understanding of the technology rather than a complete command of the technology. The larger the project, the greater the tendency for the project manager to possess an understanding rather than a command of the technology.

CHANGING VIEWS IN STAKEHOLDER MANAGEMENT

Past View	Present View
Manage existing relationships	Build relationships for the future (engagement management)
Aligned to short-term business goals	Aligned to long-term, strategic business goals
Provide ethical leadership when suited	Always provide ethical leadership
Project success is aligned to profits	Project success is aligned to client's business value
Identify profitable scope changes	Identify value-based scope changes

The present view of stakeholder management in the preceding table results from "engagement project management." In the past, whenever a sale was made to the client, the salesperson would then move on to find a new client. Salespeople viewed themselves as providers of products and/or services.

Today, salespeople view themselves as the provider of business solutions. In other words, salespeople now tell the client that "we can provide you with a solution to all of your business needs and what we want in exchange is to be treated as a strategic business partner." This benefits both the buyer and seller because:

- Not all companies (buyers) have the ability to manage complexity.

- Solution providers must learn while managing the project.

- Solution providers can bring years of history to the table.

- Solution providers have a greater understanding of cultural change, the ability to work within almost any culture, and an understanding of virtual teams.

Therefore, as a solution provider, the project manager focuses heavily on the future and a long-term partnership agreement with the client and the stakeholders. This focus is heavily oriented toward value rather than near-term profitability.

LIFE-CYCLE STAKEHOLDER MANAGEMENT

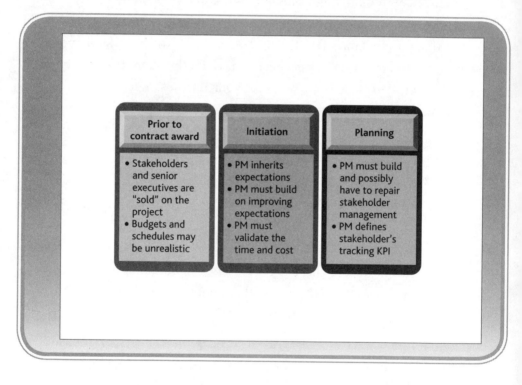

By looking at the various life-cycle stages of stakeholder management, we can see the complexities that we must endure. Prior to contract award, and before a project manager is assigned, stakeholders are sold on the project usually by someone in sales. The budget and schedule agreed to may be unrealistic since a representative from project management may not have been involved.

The project manager is brought on board at the initiation phase and inherits the promises and commitments made by the salesperson. The project manager must validate the budget and schedule, as well as building on the expectations of the various stakeholders.

In the planning stage, the realities of an unrealistic schedule and underfunded budget become apparent, and the project manager is expected to repair the damage while improving upon stakeholder management practices. The project manager must work with each stakeholder and determine what key performance indicators (KPIs) will be tracked for performance reporting.

Execution	Monitoring and controling	Closure
• PM must keep the strategic stakeholders well informed • PM must track the critical KPI for stakeholders	• PM must report the KPI variances, including high level progress reports, status reports, and end forecasting	• PM must get stakeholder buy-in on validation of deliverables and that the expectations were met

In the execution phase (or even earlier), the project manager must develop and implement a communication plan that satisfies the communication requirements of each of the stakeholders. There may be a separate communication plan for each stakeholder since there may be specific KPIs for each stakeholder.

In the monitoring and controlling phase, the project manager must prepare progress reports, status reports, and forecast reports. Once again, each report can be custom-designed for each stakeholder.

In the closure phase, which may represent the closure of a life-cycle phase or the closure of the entire project, the project manager works with the customer and the key stakeholders to validate the deliverables and that the expectations were met.

STAKEHOLDER MANAGEMENT—MACRO LEVEL

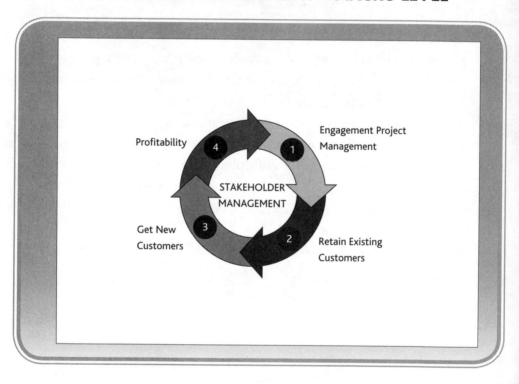

Stakeholder management can be discussed from a macro level and a micro level. The macro level begins with the engagement selling or engagement project management approach. The buyer wants someone to satisfy a business need that exists in the organization. The seller, proclaiming that they are solution providers, promises that they can provide the necessary solution and wish to establish a partnership with the buyer for future projects. The seller wants to be treated as a strategic partner rather than just a supplier.

Engagement selling, and successful project management performance, of course, allows the seller to retain existing clients as well as seeking out new customers through engagement selling. Each successful project provides the seller with the means of attracting new clients. Therefore, stakeholder management and stakeholder satisfaction is essential for the continuation of engagement selling.

STAKEHOLDER MANAGEMENT VERSUS CUSTOMER LOYALTY

- Customer loyalty and/or customer retention is not takeholder management because anyone can continuously sell products or services at a loss to support customer retention.

The final step in effective stakeholder management, shown in the previous feature, is profitability. Profitability is needed for the seller's survival. Unfortunately, appeasing stakeholders can occur the same time your company loses money by grossly underbidding a project.

In reality, customer loyalty and customer retention is not the same as stakeholder management. It is nice to have happy and loyal customers, but not without the necessary cash flow and profitability for survival. Textbooks and published papers on stakeholder management are now emphasizing that profitability must be included as part of it. The exception would be when stakeholder management is performed by nonprofit organizations.

STAKEHOLDER MANAGEMENT—MICRO LEVEL

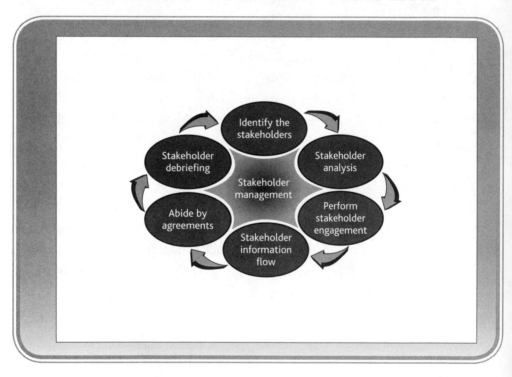

On the micro level, we can define stakeholder management using the six processes shown earlier:

- *Identify the stakeholders.* This step may require support from the project sponsor, sales, and the executive management team. Even then, there is no guarantee that all of the stakeholders will be identified.

- *Stakeholder analysis.* This requires an understanding of which stakeholders are key stakeholders that have influence, the ability and authority to make decisions, and can make or break the project. This also includes developing stakeholder management strategies, based on the results of the analysis.

- *Perform stakeholder engagements.* This step occurs when the project manager and the project team get to know the stakeholders.

- *Stakeholder information flow.* This step is the identification of the information flow network and the preparation of the necessary reports for each stakeholder.

- *Abide by agreements.* This step enforces stakeholder agreements made during the initiation and planning stages of the project.

- *Stakeholder debriefings.* This step occurs after contract closure and is to capture lessons learned and best practices for improvements on the next project involving these stakeholders.

Each of these steps is discussed in the remainder of the book.

STAKEHOLDER IDENTIFICATION

Each stakeholder is an essential piece of the project puzzle. It is not always possible to identify these supporters or adversaries without help from senior management.

Stakeholder management begins with stakeholder identification. This is easier said than done, especially if the project is multinational. Stakeholders can exist at any level of management. Corporate stakeholders are often easier to identify than political or government stakeholders.

Each stakeholder is an essential piece of the piece of the project puzzle. Stakeholders must work together and usually interact the project through the governance process. Therefore, it is essential to know which stakeholders will participate in governance and which will not.

As part of stakeholder identification, the project manager must know whether he or she has the authority or perceived status to interface with the stakeholders. Some stakeholders perceive themselves as higher stature than the project manager, and, in this case, the project sponsor may be the person to maintain interactions.

- Identified by groups
- Identified as individuals
- Identified as contributors and noncontributors to the project's success
- Identified by other factors as:
  - Authority to make decisions
  - Power and influence
  - Control of human resources
  - Source of funding
  - Technical capability
  - Others

There are several ways in which stakeholders can be identified. More than one way can be used on projects.

- *Groups.* This could include financial institutions, creditors, regulatory agencies, and the like.

- *Individuals.* This could be by name or title, such as the CIO, COO, CEO, or just the name of the contact person in the stakeholder's organization.

- *Contribution.* This could be according to financial contributor, resource contributor, or technology contributor.

- *Other factors.* This could be according to the authority to make decisions, or other such factors.

CLASSIFICATION OF STAKEHOLDERS

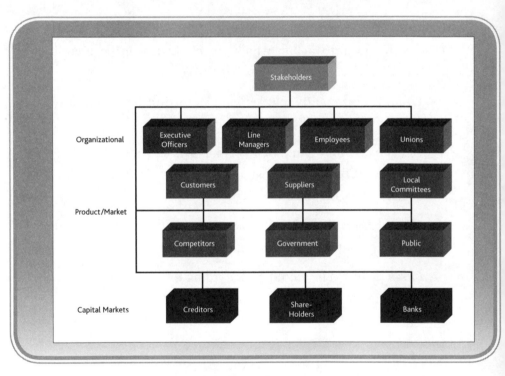

This slide shows another typical classification system for stake-holders. For simplicity sake, the stakeholders can be classified as:

- Organizational stakeholders

- Product/market stakeholders

- Capital market stakeholders

The advantage of this system is that it appears as an organizational chart, and the names of the individuals can be placed under each category.

TIERED STAKEHOLDER IDENTIFICATION

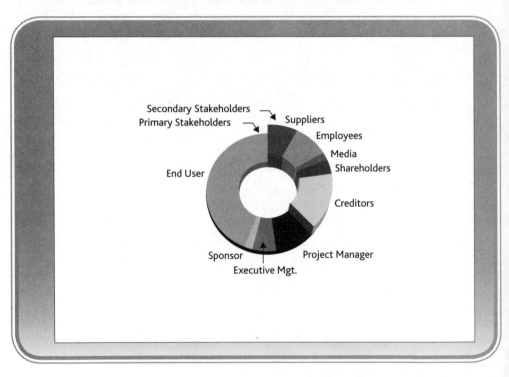

In this system of stakeholder classification, the stakeholders are identified as primary or secondary stakeholders. The primary stakeholders are highly influential stakeholders and may have voting rights. The secondary stakeholders are less influential than primarily observers. The size of stakeholder area can represent the number of stakeholders in that group.

MANAGING STAKEHOLDER EXPECTATIONS

- Each stakeholder can have different expectations concerning the project.

- Each stakeholder can have a different definition of project success.

- Each stakeholder can have different expectations on how they expect to be involved.

I t is important to understand that not all stakeholders have the same expectations on a project. Some stakeholders may want the project to succeed at any cost whereas other stakeholders may prefer to see the project fail even though they openly do not admit it. Some stakeholders view success as the completion of the project regardless of the cost overruns whereas others may define success in financial terms only.

Some stakeholders are heavily oriented toward the value they expect to see in the project, and this is the only definition of success for them. The true value may not be seen until months after the project has been completed.

Some stakeholders may view the project as their opportunity for public notice and increased stature, and therefore want to be actively involved. Others may prefer a passive involvement.

MANAGING STAKEHOLDER EXPECTATIONS: THE DESIGN OF HEALTH CARE PRODUCTS

Stakeholder	Expectations
Consumers	Must believe that the products are safe and fit for use
Stockholders	Financial interest in the selling price of the stock and the dividend
Lending Institutions	Interest rates charged for borrowing based on the present and future product revenue streams
Government	Protecting public health
Management	Protecting the company's image and reputation if any bad news occurs concerning the products
Employees	Loss of employment or income if products are a failure

Managing a project where stakeholders have different interests can be challenging. Consider a company that has a complex project to produce a new health care product. Consumers want to believe that the products developed will be safe and fit for use. Stockholders are more concerned with market share that can increase the stock's selling price and also increase the dividend. Lending institutions may be less concerned about product safety and more concerned about the revenue stream of the products such that cash flow can repay the debt.

Government agencies may have only one concern: protecting public health. Management must worry about health and product safety such that the image and reputation of the company will not be damaged if any bad news appears. Employees may provide lip service to concerns of product safety, whereas their real concern may be employment in the firm.

PERFORM STAKEHOLDER ANALYSIS

It is important to know who sits on the top of the list as the key stakeholders that can offer the greatest support throughout the project. Key stakeholders can set the direction of the project.

On large, complex projects with a multitude of stakeholders, it may be impossible for the project manager to properly cater to all of the stakeholders. Therefore, the project manager must know who the most influential stakeholders are who can provide the greatest support on the project. Typical questions to ask might include:

- Who are powerful and who are not?
- Who will have or require direct or indirect involvement?
- Who has the power to kill the project?
- What is the urgency of the deliverables?
- Who may require more or less information than others?

Not all stakeholders are equal in influence, power, or the authority to make decisions in a timely manner. It is imperative for the project manager to know who sits on the top of the list.

Finally, it is important to remember that stakeholders can change over the life of a project, especially if it is a long-term project. Also, the importance of certain stakeholders can change over the life of a project and in each life-cycle phase. The stakeholder list is therefore an organic document subject to change.

STAKEHOLDER MAPPING

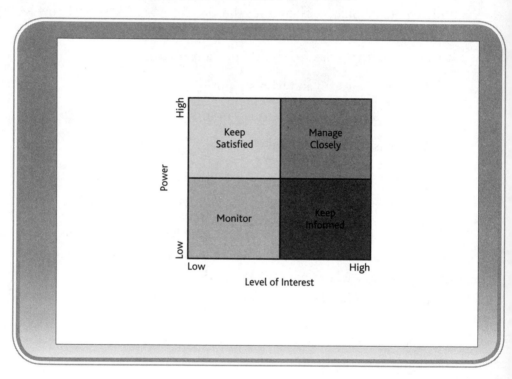

Stakeholder mapping is most frequently displayed on a grid comparing their power and their level of interest.

- *Manage closely.* These are high-power, interested people who can make or break your project. You must put forth the greatest effort to satisfy them. Be aware that there are factors that can cause them to change quadrants rapidly.

- *Keep satisfied.* These are high-power, less interested people who can also make or break your project. You must put forth some effort to satisfy them but not with excessive detail that can lead to boredom and total disinterest. They may not get involved until the end of the project approaches.

- *Keep informed.* These are people who have limited power but are keenly interested in the project. They can function as an early warning system of approaching problems and may be technically astute to assist with some technical issues. These are the stakeholders who often provide hidden opportunities.

- *Monitor only.* These are people who have limited power and may not be interested in the project unless a disaster occurs. Provide them with some information but not with so much detail that they will become disinterested or bored.

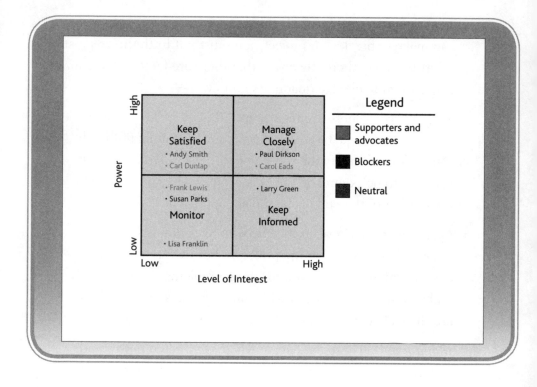

Stakeholder mapping can also take place with the names of the people placed in the appropriate quadrants such as seen in the above figure. The names can be color-coded to identify the supporters or advocates, blockers, and those that appear neutral. It is important to note that supporters, blockers, and neutral positions can appear in any quadrant and that names can move from quadrant to quadrant based on changes that occur in each life-cycle phase.

This four-quadrant technique for categorizing stakeholders is only one of many. Another technique is the Stakeholder Salience Model[1] that examines three stakeholder attributes—power, legitimacy, and urgency—rather than the two attributes of the four-quadrant model. In fact, some organizations have detailed spreadsheets, which compare 10 or more weighted stakeholder attributes, the results of which are used for stakeholder management categorization. Each organization needs to figure out which technique works best for them. They may even have a variety of available techniques for project management teams to use.

---

[1]Mitchell, R., Agle, B., Wood, D., *Toward a Theory of Stakeholder Identification and Salience: Defining the Principle of Who And What Really Counts*, Academy of Management Review 1997, Vol. 22, No. 4, 853–886.

KEY STAKEHOLDERS

- You must win their support.

- They may be able to interpret and influence the internal and external environments.

- They can identify enterprise environmental factors.

- They may be able to improve your organizational process assets.

- They may be able to provide additional resources.

The larger the project, the more important it becomes to know who is and is not an influential or key stakeholder. Although you must win the support of all stakeholders, or at least try to do so, the key stakeholders come first.

Key stakeholders may be able to provide the project manager with assistance with the identification of enterprise environmental factors that can impact the project. This could include forecasting on the host country's political and economic conditions, the identification of potential sources for additional funding, and other such issues.

In some cases, the stakeholders may have software tools that can supplement the project manager's available organizational process assets.

UNIMPORTANT STAKEHOLDERS

- You must win the support of the neutral, unimportant, or skeptical stakeholders as you would key stakeholders. This may be a challenge.

- Unimportant stakeholders may become important as the end of the project nears.

Thus far, we have discussed the importance of winning over the key or influential stakeholders. There is also a valid argument for winning over the stakeholders that are considered to be unimportant. While some stakeholders may appear to be unimportant, that can change rapidly. For example, an unimportant stakeholder suddenly discovers that a scope change is about to be approved and that scope change can seriously impact the unimportant stakeholder, perhaps politically. Now, the unimportant stakeholder (originally deemed so for apparent lack of concern about the project) becomes a key stakeholder.

Another example occurs on longer-term projects where stakeholders may change over time perhaps because of politics, promotions, retirements, or reassignments. The new stakeholder may suddenly want to be an important stakeholder whereas his or her predecessor was more of an observer.

Finally, stakeholders may be relatively quiet in one life cycle because of limited involvement but become more active in other life cycles where they must participate. The same may hold true for people that are key stakeholders in early life-cycle phases and just observers in later phases.

The project team must know who the stakeholders are. The team must also be able to determine which stakeholders are critical stakeholders at specific points in time.

PERFORM STAKEHOLDER ENGAGEMENTS

It is important to get to know your stakeholders as quickly as possible. Searching out the critical data requires diplomacy and effective communications.

Stakeholder engagement occurs when you physically meet with the stakeholders and determine their needs and expectations:

- Understand them and their expectations.

- Understand their needs.

- Value their opinions.

- Find ways to win their support on a continuous basis.

- Identify any stakeholder problems early on that can influence the project.

Even though stakeholder engagement follows stakeholder identification, it is often through stakeholder engagement that we determine which stakeholders are supporters, advocates, neutral, or opponents. This may also be viewed as the first step in building a trusting relationship between the project manager and the stakeholders.

DEFINING KEY PERFORMANCE INDICATORS (KPIs)

- We must establish KPI milestones.

- Milestones must be value driven.

- Each stakeholder may ask for a different set of KPIs; this is a necessity to maintain their interest.

- Obtaining KPI agreement from all stakeholders may be difficult.

- We may need multiple dashboards; this may be costly but necessary.

As part of stakeholder engagements, it is necessary for the project manager to understand each stakeholder's interests. One of the ways to accomplish this is to ask the stakeholders (usually the key stakeholders) what information they would like to see in performance reports. This information will help identify the KPIs needed to service this stakeholder.

Each stakeholder may have a different set of KPI interests. This then becomes a costly endeavor for the project manager to maintain multiple KPI tracking and reporting flows, but it is a necessity for successful stakeholder management. Getting all of the stakeholders to agree on a uniform set of KPI reports and dashboards may be almost impossible.

PRIORITIZING STAKEHOLDERS' NEEDS

	Stakeholders				
	Customers	Shareholders	Government	Management	Employees
Product Quality	A	C	B	B	B
Product Safety	A	C	A	C	C
Product Features	A	C	C	C	B
Product Cost	B	A	C	A	C
Delivery Date	A	B	C	A	A

A = High Stakeholder Importance
B = Somewhat Important to Stakeholder
C = Low Stakeholder Importance

The preceding table shows the complexities in getting stakeholder agreement. Each stakeholder will have his or her own issues and challenges. In a table such as this, we can prioritize the importance of each issue to each stakeholder and then try to determine the minimum number of KPIs to track and report on the high-priority issues.

In reality, the list of the number of issues facing all of the stakeholders could be quite long.

STAKEHOLDER INFORMATION FLOW

Stakeholder information flow is the determination of who wants what information when and in what format. The Internet will most likely serve as the primary mechanism to do this.

There must be an agreement on what information is needed for each stakeholder, when the information is needed, and in what format the information will be presented. Some stakeholders may want a daily or weekly information flow, whereas others may be happy with monthly data. For the most part, the information will be provided via the Internet.

Project managers should use a communications matrix to carefully lay out planned stakeholder communications. Information in this matrix might include the definition or title of the communication (e.g., status report, risk register), the originator, the intended recipients, the medium to be used, rules for access, and frequency of publication or updates.

Project managers astute in dashboard design will prepare dashboards for specific stakeholders. Some dashboards will contain real-time data, whereas others may be updated monthly. Factors to be considered in dashboard information flow include:

- Colors

- Positioning

- Brightness

- Orientation

- Saturation

- Size

- Texture

- Shape

Some rules exist for dashboard design and layout:

- Rules for selecting the right artwork

- Rules for artwork placement

- Rules for color selection

- Rules for accuracy of information

Examples of rules include:

- Must select the correct metaphor (i.e., gauges cannot show trends; pretty artwork can distract users from critical information)

- Speed of perception (i.e., upper left corner is more critical than lower right corner)

- Visualization (i.e., easy to read and understand)

- Aesthetics (i.e., pleasing to the eye)

VIRTUAL TEAMS

It is important to understand that most complex projects will be using some virtual teams. Once again, we rely on the Internet for effective communications.

The more complex the project, the greater the need for virtual teams. Virtual teams thrive on effective communications. And if the information provided in the performance reports and the tracking of the KPIs are accurate, trust will build up among the project team and the various stakeholders.

Since virtual teams may be remotely located from where the work is taking place, they must rely heavily on communications and be made to feel that the information they receive is true. As part of stakeholder information flow for both virtual and nonvirtual teams, the project manager must:

- Prepare a communication plan that identifies the reporting needs of each stakeholder (amount of information, level of detail, etc.).

- Identify stakeholder-specific KPIs.

- Identify communication protocols.

- Identify any proprietary information requirements or security needs.

- Continuously focus on the value and benefits at completion of the project.

MEASURING KPIs

Reporting KPI data begins with KPI measurements. Each KPI can be unique to a particular stakeholder, and the project team may be inexperienced with the KPI such that accurate measurement cannot take place.

Previously, we discussed the complexities of determining the KPIs for each stakeholder. Some issues that need to be addressed include:

- The potential difficulty in getting customer and stakeholder agreement on the KPIs

- Determining if the KPI data is in the system or needs to be collected

- Determining the cost, complexity, and timing for obtaining the data

- Considering the risks of information system changes and/or obsolescence that can impact KPI data collection over the life of the project

KPIs have to be measurable, but some KPI information may be difficult to quantify. For example, customer satisfaction, goodwill, and reputation may be important to some stakeholders, but they may be difficult to quantify. Some KPI data may need to be measured in qualitative terms rather than quantitative terms.

Some KPI data, such as the quality of the deliverables, cannot be tracked until the deliverable is completed. Eckerson has prepared a list identifying the criteria for a KPI such that measurement may be possible:

- Aligned with a strategy or objectives

- Owned by a group accountable for the outcome

- Predictive indicator of the future

- Actionable by allowing for improvements

- Few in number

- Easy to understand

- Balanced and linked (reinforce each other)

- Trigger changes

- Standardized (appear in dashboards)

- Context driven (have targets and thresholds)

- Reinforced with incentives

- Relevant (periodically reviewed and refreshed)[2]

---

[2] Wayne W. Eckerson, *Performance Dashboards: Measuring, Monitoring and Managing Your Business*. Hoboken, NJ: John Wiley & Sons, 2006, p. 201.

REPORTING KPI DATA

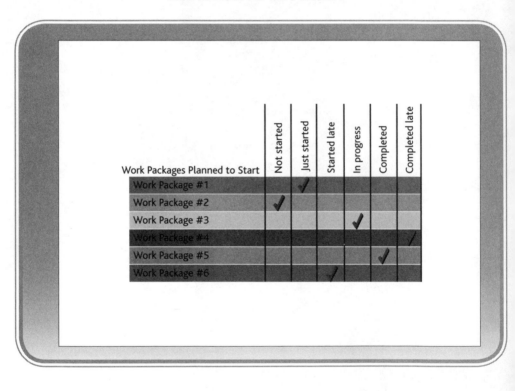

Work Packages Planned to Start	Not started	Just started	Started late	In progress	Completed	Completed late
Work Package #1		✓				
Work Package #2	✓					
Work Package #3				✓		
Work Package #4						✓
Work Package #5					✓	
Work Package #6			✓			

There are a variety of techniques for KPI reporting ranging from written reports to dashboards. The preceding feature represents information for a stakeholder who is interested in work package progress. The information can be updated on a dashboard using real-time data by simply inserting a check mark in the appropriate location.

SUMMARIZED KPI MILESTONES

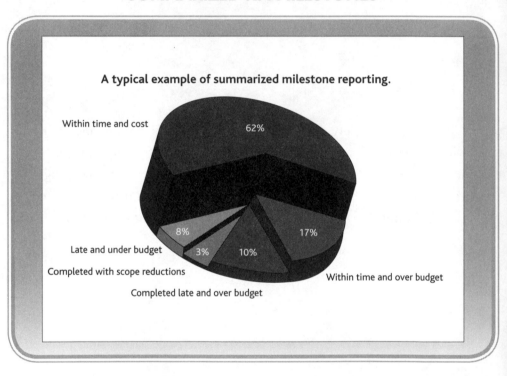

A typical example of summarized milestone reporting.

Within time and cost
62%
Late and under budget 8%
Completed with scope reductions 3%
Completed late and over budget 10%
Within time and over budget 17%

The preceding feature might be representative of a summary KPI report. For example, 17 percent of the milestones thus far were completed within time but over budget. This information can also be used in dashboard reporting and updated using real-time data. This KPI would be appropriate for stakeholders that are mainly observers and just interested in summary information on the project's performance thus far.

With real-time data streams updating these KPIs, it is also possible to click on each sector in the figure on the dashboard and get a more detailed breakdown of specifically which milestones are in each sector. Real-time data stream reporting allows dashboards to be designed such that they can satisfy the needs of both the primary and secondary stakeholders, thus reducing the time and effort for managing some stakeholders on an individual basis.

STAKEHOLDER COMMUNICATIONS

Because each stakeholder may have different needs, project managers may need to prepare a variety of customized reports, and this can be tedious and expensive.

The need for effective stakeholder communications is clear:

- Communicating with stakeholders on a regular basis is a necessity.

- By knowing the stakeholders, you may be able to anticipate their actions.

- Effective stakeholder communications builds trust.

- Virtual teams thrive on effective stakeholder communications.

- Although we classify stakeholders by groups or organizations, we still communicate with people.

- Ineffective stakeholder communications can cause a supporter to become a blocker.

PROJECT REVIEW MEETINGS

- Project review meetings with stakeholders are not the same as project team meetings.
- Each stakeholder review meeting has its own characteristics.
- At each meeting, stakeholders must be convinced that:

  - Project management is working well and as planned.
  - The expected value will be there at the end.

There are two types of project review meetings: those with the project team and those with other stakeholders. Stakeholder meetings have their own characteristics. The items discussed in the stakeholder meetings include:

- A review of the stakeholder-specific KPI information

- A discussion of how well project management is working

- Forecasts for the time, cost, benefits, and value expected at the end of the project

These meetings also are used to resolve problems. Project managers must find solutions to problems such that multiple stakeholders are satisfied simultaneously, or else the project manager will be overwhelmed in meetings. It may be necessary to have multiple stakeholders attending the same meeting if there are common issues to be resolved.

STAKEHOLDER SCOPE CHANGE REQUESTS

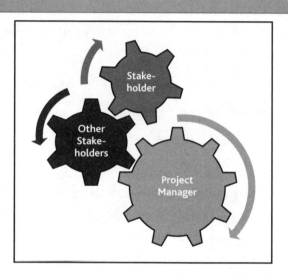

The impact of a small change requested by one stakeholder can have a major impact on other stakeholders and the project manager.

The opportunities for scope changes abound on every project. On large, complex projects, there may be one or more individuals assigned to the project office simply for the management of scope changes. Scope changes can be approved and implemented incrementally as the project progresses, or all scope changes can be withheld until after the project is completed and then implemented as an enhancement project.

On complex projects with a large number of stakeholders, the advantages of a scope change from the perspective of one stakeholder could be viewed by another stakeholder as a disadvantage. Stakeholders often recommend and approve scope changes based on what they perceive is in their own personal interest, while neglecting the interests of those around them.

In the above figure, assume that the stakeholder in the smaller wheel recommends what they consider to be a small scope change. Notice how quickly the second wheel must turn to keep up with the smaller wheel. The largest wheel must move even faster to keep up with the two smaller wheels. While this may be an exaggeration, it does show that the impact of a scope change on one stakeholder can have a variety of effects on other stakeholders.

LINEAR THINKING

■ Traditional enterprise project management (EPM) methodologies are based on linear thinking and do not necessarily
allow for effective stakeholder communication.

■ This happens even if the methodology has templates
for stakeholder identification and how to work with
stakeholders.

Most companies today have EPM methodologies that focus on linear thinking. All project work follows well-established life-cycle phases. The project manager also has forms, guidelines, templates, and checklists for each phase. This linear thinking may not be appropriate for many of today's complex projects.

Today's complex projects may require a fluid or adaptive methodology that can be custom-designed to be applied differently for each client and stakeholder. There may be different tools for each client. Therefore, project managers may need to use outside-of-the-box thinking to give each stakeholder the attention he or she expects.

ENFORCING STAKEHOLDER AGREEMENTS

It is important to stay on top of all stakeholder agreements. Not all stakeholders may abide by their original agreements, mainly due to politics, and changes in stakeholders may require new agreements.

Part of the process of stakeholder engagement involves the establishment of agreements between the individual stakeholders and the project manager and among other stakeholders as well. These agreements must be enforced throughout the project. The project manager must identify:

- Any and all agreements among stakeholders (i.e., funding limitations, sharing of information, approval cycle for changes, etc.)

- How politics may change stakeholder agreements

- Which stakeholders may be replaced during the project (i.e., retirement, promotion, change of assignment, politics, etc.)

The project manager must be prepared for the fact that not all agreements will be honored.

STAKEHOLDER DEBRIEFING SESSIONS

- This is a project closeout review session.
- Identify what went well.
- Identify areas of improvement for future contracts with this stakeholder.
- Attendance can include:
  - Engineering
  - Manufacturing
  - Sales/marketing
  - Senior management
  - The project management office

These types of stakeholder debriefing sessions occur at the closure of the project, usually after contractual closure. Some companies have a life-cycle phase after contractual closure entitled Customer Satisfaction Management. With engagement project management or engagement selling, you want to build a strategic partnership relationship with your client and the stakeholders. The intent of these sessions is to determine:

- What did we do well on this project performance-wise and also with regard to stakeholder management?

- What did we do poorly on this project performance-wise and with regard to stakeholder management?

- What are the areas for improvement?

Attendance at the meeting is not restricted to just the project manager and the stakeholders. Attendance can include the sales team that sold the stakeholders on the project, the sponsors that interfaced with the stakeholders, and senior management hoping for additional work.

SATISFACTION MANAGEMENT SURVEY FACTORS

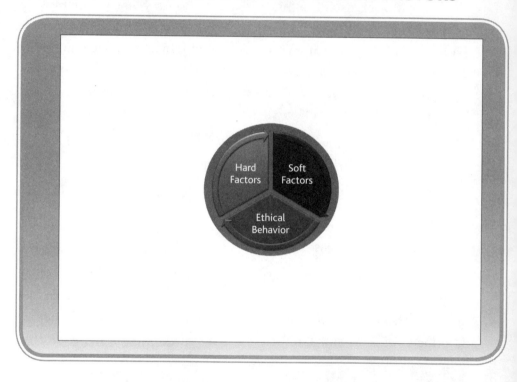

Satisfaction management surveys can occur throughout the project as well as at the end. There are three areas of interest in these surveys:

- Hard satisfaction management data that includes performance data like KPI data and milestone/deliverable measurements

- Soft satisfaction management data such as effective stake-holder interfacing and communications

- Ethical behavior data

Some companies use templates for this with boxes to be checked, ranging from "completely unsatisfied" to "completely satisfied." When boxes are checked that are not in the "completely satisfied" or just "satisfied" categories, guidelines may exist on what to do to next to proceed to a higher level of satisfaction.

COMPLEX PROJECT MANAGEMENT SKILLS

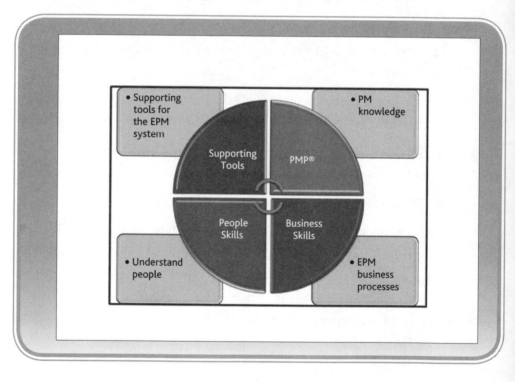

The *PMBOK® Guide* identifies the nine areas of knowledge that project managers should understand. However, the relative importance of each of these areas can change from project to project as well as from life-cycle phase to life-cycle phase. For complex projects, there may be other knowledge requirements that are not specifically discussed in the *PMBOK® Guide*.

The preceding feature shows the four areas where complex projects may differ from traditional projects. Business skills are essential for complex projects, especially when dealing with a multitude of stakeholders. Companies that compete in the global marketplace as solution providers have significantly more tools available for the project managers, and the majority of these tools are business-oriented tools.

People skills on complex projects may need to emphasize the following:

- Presentation skills
- Coping under stress
- Managing virtual teams
- Conflict resolution skills
- Stress management
- Mentorship
- Counseling and facilitation
- Decision-making skills
- How to conduct effective meetings
- Dashboard design techniques

THREE CRITICAL FACTORS FOR SUCCESSFUL
STAKEHOLDER MANAGEMENT

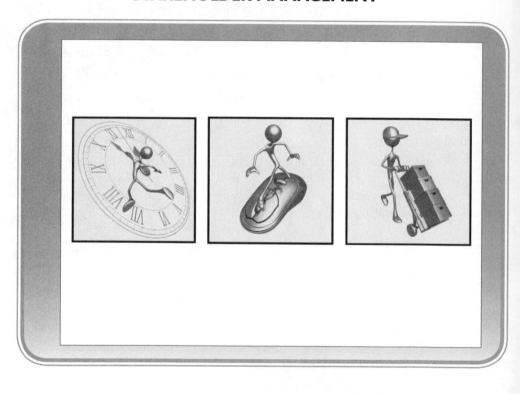

There are three additional critical factors that must be considered for successful stakeholder management:

- Effective stakeholder management takes time. It may be necessary to share this responsibility with sponsor, executives, and members of the project team.

- Based on the number of stakeholders, it may not be possible to address their concerns face to face. You must maximize your ability to communicate via the Internet. This is also important when managing virtual teams.

- Regardless of the number of stakeholders, documentation on the working relationships with the stakeholders must be archived. This is critical for success on future projects.

Effective stakeholder management can be the difference between an outstanding success and a terrible failure.

SUCCESSFUL STAKEHOLDER MANAGEMENT

■ Successful stakeholder management will result in agreement on:

- ■ Vision

- ■ Objectives

- ■ Targeted or end value

- ■ Each one's support level

- ■ How stakeholders will interact among each other

- ■ How performance will be tracked (i.e., KPIs)

- ■ The depth and frequency of reporting

- ■ The scope change control process and approvals

Successful stakeholder management can result in agreements on the items in the preceding feature. The resulting benefits will be:

- Better decision making and in a more timely manner

- Better control of scope changes; prevention of unnecessary changes

- Follow-on work from stakeholders

- End-user satisfaction and loyalty

- Minimizing the impact that politics can have on your project

FAILURES IN STAKEHOLDER MANAGEMENT

Sometimes, regardless of how hard we try, we will fail at stakeholder management. Typical reasons include:

- Inviting stakeholders to participate too early, thus encouraging scope changes and costly delays

- Inviting stakeholders to participate too late such that their views cannot be considered without costly delays

- Inviting the wrong stakeholders to participate in critical decisions, thus leading to unnecessary changes and criticism by key stakeholders

- Key stakeholders become disinterested in the project

- Key stakeholders are impatient with the lack of progress

- Allowing the key stakeholders to believe that their contributions are meaningless

- Managing the project with an unethical leadership style or interfacing with the stakeholders in an unethical manner

FINAL THOUGHTS

In this book, we have attempted to address the implications of project complexity on the good practices that are described in the *Guide to the Project Management Body of Knowledge*—Fourth Edition®. We have examined the increased importance of the role of stakeholders in projects of advanced complexity. In fact, the changing roles of the stakeholders and their varying degrees of involvement in the project add to the project's complexity. We have also seen how the management of project risk takes on added importance. Many of our observations about specific knowledge areas of the *PMBOK*® *Guide* deal with the permutations of increased risk attached to those areas. Finally, we've noted the need for new tools, lifecycles, and techniques demanded by increased project complexity.

We must keep in mind that project management is an ever evolving practice. The *PMBOK*® *Guide*, now in its fourth edition, makes clear that its contents describe "good practices" for managing projects—not necessarily the "best practices" for any specific project or industry group. Those will ultimately need to be defined by the organizations that are managing and performing the work of the project. We note that there are still rival philosophies for certification of project managers (PMP from PMI, Prince2 from United Kingdom Office of Government Commerce, and IPMA levels C, B, and A). In addition, as this book is going to press, the International Organization for Standardization (ISO) is floating a Committee Draft (CD) of Standard for Project Management ISO 21500 within its participating and decision-making countries and parties.

The area of project complexity is likewise slowly evolving, and this book merely scratches the surface of that evolution. Its impact is currently being addressed by major project management professional organizations. Within a few months of this book's publication, the International Centre for Complex Project Management is scheduled to begin discussing the development of standards. Perhaps we will even get agreement on defining what exactly makes a project complex. Much of the more sophisticated thinking around project complexity revolves around complexity theory itself. The reader is encouraged to continue examining areas associated with complexity theory—edge of chaos, network theory, landscapes, adaptability, and emergence, just to name a few. Projects are essentially endeavors of people and organizations to meet specific needs through available technologies in given environments. And as those people, organizations, technologies, and environments change, so too will the processes, tools, and techniques that are used to manage them.

INDEX

389